Eva Benz

The CO2 Allowance Price in the European Emissions Trading Scheme

Eva Benz

The CO2 Allowance Price in the European Emissions Trading Scheme

An Empirical, Experimental, and Theoretical Study

Südwestdeutscher Verlag für Hochschulschriften

Impressum/Imprint (nur für Deutschland/ only for Germany)
Bibliografische Information der Deutschen Nationalbibliothek: Die Deutsche Nationalbibliothek verzeichnet diese Publikation in der Deutschen Nationalbibliografie; detaillierte bibliografische Daten sind im Internet über http://dnb.d-nb.de abrufbar.
Alle in diesem Buch genannten Marken und Produktnamen unterliegen warenzeichen-, marken- oder patentrechtlichem Schutz bzw. sind Warenzeichen oder eingetragene Warenzeichen der jeweiligen Inhaber. Die Wiedergabe von Marken, Produktnamen, Gebrauchsnamen, Handelsnamen, Warenbezeichnungen u.s.w. in diesem Werk berechtigt auch ohne besondere Kennzeichnung nicht zu der Annahme, dass solche Namen im Sinne der Warenzeichen- und Markenschutzgesetzgebung als frei zu betrachten wären und daher von jedermann benutzt werden dürften.

Verlag: Südwestdeutscher Verlag für Hochschulschriften Aktiengesellschaft & Co. KG
Dudweiler Landstr. 99, 66123 Saarbrücken, Deutschland
Telefon +49 681 37 20 271-1, Telefax +49 681 37 20 271-0, Email: info@svh-verlag.de
Zugl.: Bonn, Bonn Graduate School of Economics, Diss., 2008

Herstellung in Deutschland:
Schaltungsdienst Lange o.H.G., Berlin
Books on Demand GmbH, Norderstedt
Reha GmbH, Saarbrücken
Amazon Distribution GmbH, Leipzig
ISBN: 978-3-8381-0553-6

Imprint (only for USA, GB)
Bibliographic information published by the Deutsche Nationalbibliothek: The Deutsche Nationalbibliothek lists this publication in the Deutsche Nationalbibliografie; detailed bibliographic data are available in the Internet at http://dnb.d-nb.de.
Any brand names and product names mentioned in this book are subject to trademark, brand or patent protection and are trademarks or registered trademarks of their respective holders. The use of brand names, product names, common names, trade names, product descriptions etc. even without a particular marking in this works is in no way to be construed to mean that such names may be regarded as unrestricted in respect of trademark and brand protection legislation and could thus be used by anyone.

Publisher:
Südwestdeutscher Verlag für Hochschulschriften Aktiengesellschaft & Co. KG
Dudweiler Landstr. 99, 66123 Saarbrücken, Germany
Phone +49 681 37 20 271-1, Fax +49 681 37 20 271-0, Email: info@svh-verlag.de

Copyright © 2009 by the author and Südwestdeutscher Verlag für Hochschulschriften Aktiengesellschaft & Co. KG and licensors
All rights reserved. Saarbrücken 2009

Printed in the U.S.A.
Printed in the U.K. by (see last page)
ISBN: 978-3-8381-0553-6

Contents

Introduction   1

1   Modeling the Price Dynamics of CO_2 Emission Allowances   7
    1.1   Introduction   7
    1.2   Market Mechanism and Instruments   10
        1.2.1   The EU ETS and Classification of Emission Allowances   10
        1.2.2   Price Determinants of CO_2 Emission Allowances   11
    1.3   Modeling the Price Dynamics of CO_2 Emission Allowances   13
        1.3.1   GARCH Models   13
        1.3.2   Regime-Switching Models   14
    1.4   Empirical Results   17
        1.4.1   The Data   17
        1.4.2   Analyzing Logreturns of CO_2 Allowances   21
        1.4.3   Time series Models   22
        1.4.4   In-Sample Results   23
        1.4.5   Forecasting Results   31
        1.4.6   Comparison with results from other papers   36
    1.5   Summary and Conclusion   38

2   Liquidity and Price Discovery in the CO_2 Futures Market: An Intraday Analysis   41
    2.1   Introduction   41
    2.2   Market Structure and Data   44

		2.2.1 Market Structure of Carbon Exchanges	44
	2.3	Market Structure and Data	44
		2.3.1 Institutional Background	44
		2.3.2 Market Structure of Carbon Exchanges	45
		2.3.3 Data Set and Summary Statistics	49
	2.4	Spread Analysis	54
		2.4.1 Methodology	56
		2.4.2 Estimation Approach	58
		2.4.3 Estimation Results	58
	2.5	Price Discovery	64
		2.5.1 Methodology	65
		2.5.2 Estimation Results	69
	2.6	Conclusion	75
3	The Initial Allocation of CO_2 Emission Allowances: An Experimental Study		79
	3.1	Introduction	79
	3.2	Initial Allocation Rules and Criteria	82
	3.3	Related Literature	85
	3.4	Experimental Design	86
		3.4.1 Emissions Trading Game	87
		3.4.2 Treatments	88
		3.4.3 Subjects' Characteristics	90
		3.4.4 Organization of the Experiment	92
	3.5	Theoretical Considerations	93
	3.6	Experimental Results	96
		3.6.1 Prices, Volumes, and Allocation Efficiency	96
		3.6.2 Bidding Strategies	101
	3.7	Conclusion	106

CONTENTS

Appendices		109
A	Appendix to Chapter 2	109
	A.1 Components of Estimated Half Spreads	109
B	Appendix to Chapter 3	111
	B.1 Design of Dynamic Uniform Double Auction	111
	B.2 Two-Stage Model	111
	B.3 Instructions	118

Bibliography 125

List of Tables

1.1	Summary statistics for the EUA logreturns	21
1.2	Parameter estimates of the AR(1)-GARCH(1,1) model	26
1.3	Estimation results with the two-state regime-switching model	27
1.4	Model evaluation by the log-likelihood and information criteria	31
1.5	Point forecast results	32
1.6	Results for Kolmogorov-Smirnov and Kuiper statistics.	37
2.1	Overall trading volume of the EUA spot and futures market in Phase 1	46
2.2	Trading volumes (without OTC) of EUA futures with expiry in December	50
2.3	Estimated half spreads for the common sample periods	59
2.4	Estimated half spreads for the most liquid years and its quarters	61
2.5	Results for stationarity tests	71
2.6	Estimation results of the VECM for Phase 1	72
2.7	Estimation results of the VECM for a restricted period, daily transaction frequencies, and trading volume	74
3.1	Basic characteristics for all treatments	89
3.2	Total allocated quantities via grandfathering and auctioning	89
3.3	Subjects' characteristics in the treatments	91
3.4	Organization of the experiment	92
3.5	Evaluation of initial allocation rules with respect to the three criteria	95
3.6	Sequence of reference prices in the treatments	95

3.7	Average observed prices and deviations from the cost minimum	97
3.8	Average observed and optimal volumes	99
3.9	Degree of cost-efficiency	100
3.10	Percentage of dropouts in the auction that are consistent with MAC and p^*	102
3.11	Distribution of submitted offer prices	105
A.12	Components of estimated half spreads	110
B.13	Example of myopic bidding behavior	117

List of Figures

1.1	Daily EUA Prices from August 27, 2003 - December 29, 2006	18
1.2	Daily EUA logreturns from January 3, 2005 - December 29, 2006	20
1.3	Empirical distribution and Gaussian fit to EUA logreturns	22
1.4	Innovations, conditional standard deviations and logreturns of AR(1)-GARCH(1,1)	25
1.5	Logreturns of EUA prices and probability of being in the spike regime I . .	28
1.6	Logreturns of EUA prices and probability of being in the spike regime II .	30
1.7	One-day ahead forecast results	34
1.8	Logreturns of EUA prices and predicted 95%-confidence intervals	35
2.1	EUA futures prices in Phase 1	51
2.2	Monthly transaction frequencies in Phase 1	52
2.3	Monthly average return standard deviation	53
2.4	Intraday pattern of estimated half spreads	63
3.1	Average observed prices for each treatment	97
3.2	Histograms of dropout prices in the auction	103
3.3	Histograms of first selling offers in the auction	104

Introduction

According to the common position of the European Council, large installations from the energy industry and other carbon-intensive industries are part of an EU-wide greenhouse gas (GHG) emissions trading scheme (EU ETS) that formally has entered into operation in January 2005. So far it is the world's largest GHG emissions trading system covering over 10,000 installations in the energy and industrial sectors that are collectively responsible for about 50% of Europe's CO_2 emissions and 40% of its total GHG emissions. It is considered as the cornerstone of the European Climate Change Programme and is expected to help achieving the EU's obligations under the United Nations Framework Convention on Climate Change and the Kyoto Protocol in a cost-effective way. Under the Kyoto Protocol the EU has committed to reducing GHG emissions by 8% compared to the 1990 level by the years 2008-2012.

The EU ETS is made up of consecutive trading periods. The first trading period (Phase 1) served as a pilot phase and covers the years 2005-2007 while the second trading period from 2008-2012 constitutes the Kyoto commitment period (Phase 2). Plans for the post Kyoto trading period 2013-2020 (Phase 3) became more concrete after the United Nations summit in Bali in December 2007. Besides, in January 2008 the European Commission has agreed on a so called "Climate and Energy Package", which makes first regulatory suggestions and improvements for the continuation of action against climate change in the EU. In the first two phases only CO_2 emissions have been affected and for Phase 3 the European Commission intends to include the GHG nitrous oxide and perfluorocarbons.

The EU ETS is organized as a cap-and-trade scheme where participating firms have to reduce the amount of emitted CO_2 and annually demonstrate that their level of European Union CO_2 allowances (EUAs) corresponds to their actual emissions. Every year, at the end of February, a certain amount of EUAs is allocated to the compliant firms for the current trading year according to National Allocation Plans (NAPs). On April 30 of the following year, firms have to deliver the required EUAs to the national surveillance authorities according to their actual emissions volume. Not handing in the required amount of EUAs is fined with an extra fee of Euro 40 (Euro 100) per missing EUA in the Phase 1 (Phase 2) additional to delivering the missing amount of EUAs. One allowance covers the equivalent of one ton of CO_2 emissions. Companies being able to keep emissions below their allocation level are free to sell excess allowances in the market. Firms which need additional allowances to comply with their output levels have the choice to either invest in emissions-reducing technologies, to switch to less emissions-intensive production technologies or, if marginal abatement costs are higher than the market price of EUAs, to buy EUAs on the European CO_2 market. Within Phase 1 and Phase 2 surplus allowances could be transferred for use during the following year (banking). Banking between Phase 1 and Phase 2 was forbidden by most of the countries. Only France and Poland allowed for restricted banking. As allocation always takes place in February, borrowing of EUAs from the future year is indirectly possible as the compliance date for the preceding year is April 30. However, it was not possible to borrow EUAs between 2007 and 2008.

Allowance trading has primarily been applied in the US, where it has become a crucial policy instrument to address air as well as nutrient pollution in water bodies at federal and state level. The majority of publications about tradable allowances assesses the Acid Rain Program of the US Environmental Protection Agency (EPA), where operators of power plants have been trading sulphur dioxide (SO_2) allowances since 1992.

The theory behind using emissions trading as environmental policy instrument dates back to Montgomery (1972). Using a static model for a perfect market with pollution certificates, he was the first to show that there exists a minimum cost equilibrium for companies facing a given environmental target. Rubin (1996) extended his model to a dynamic setting with intertemporal transfer of the assets and confirmed the existence of a cost efficient solution.

However, Kling and Rubin (1997) showed that in this framework the equilibrium does not coincide with the welfare maximizing first-best solution. According to Tietenberg (1990) reaching the cost minimal equilibrium requires that the marginal cost for abating emissions of all companies equals the market price of emission allowances. Thus, from a theoretical perspective introducing tradable permits is considered as a cost efficient instrument for reaching environmental targets where the initial allocation of allowances does not matter. This dissertation sheds light on the price of CO_2 emission allowances by using empirical and experimental analysis as well as theoretical models. In particular, throughout this work I examine the EUA price pattern that has evolved in Phase 1. For this purpose the first two chapters provide an empirical analysis of EUA prices, with Chapter 1 studying daily spot prices and Chapter 2 high frequency data for futures prices. Finally in the last chapter, I analyze CO_2 allowance prices that evolve under different policy and market behavior scenarios in order to give recommendations for a viable trading scheme in Phase 3. I present a laboratory experiment that is concerned with the annual initial allocation process of EUAs to the participating firms where the focus is on using auctioning as (part of the) allocation mechanism. The hypotheses for the analysis of the experimental data are derived from a theoretical model that is in line with the experimental setup.

In the next paragraphs, I present the main topics addressed in my thesis in more detail. The aim of the first chapter[1] is to provide an in-sample analysis of the short-term spot price behavior of EUAs focusing on the price dynamics and changes in the volatility of the underlying stochastic price process. As CO_2 allowances display similarities to operational materials or commodities, I adopt commodity pricing models such as Markov switching and AR-GARCH models for stochastic modeling. Additionally, in this chapter I concentrate on the out-of-sample performance of the models with respect to forecasting. In particular, I evaluate price, volatility and density forecasts for the different approaches. Both analyses are of interest to risk managers and traders who constantly have to hedge their positions against unexpected carbon price fluctuations. I show that the best in-sample fit to the data is provided by a regime-switching model with an autoregressive process in the base

---
[1]This chapter is based on the paper by Benz and Trück (2008).

regime and a normal distribution for the spike regime. The results for the GARCH and normal mixture regime-switching model are slightly worse. The comparison of one-day-ahead point and density forecasts of the different pricing models yields that for the one-day ahead density forecasts, the AR-GARCH and regime-switching model clearly outperform the models with constant variance.

Chapter 2^2 uses the end of Phase 1 in December 2007 to give a comprehensive overview of the European carbon market development. Having access to intraday transactions data, I additionally investigate this recent market from a microstructure angle. Since almost all trading takes place in the futures markets, I focus on EUA futures price data supplied by ECX and Nord Pool, the two most liquid European trading exchanges for EUA futures. I compare two microstructure issues that are of high relevance to potential traders: liquidity and price discovery. With respect to liquidity, I study overall trading volumes as well as the development of trading frequencies across exchanges and estimate traded bid-ask spreads. To analyze relative price discovery on both exchanges I use a vector error correction framework that builds on the cointegration relationship between transaction price series. To quantify the two markets' relative contributions to the price discovery process I apply two different measures. The results reveal that estimated transaction spreads markedly decrease on both exchanges over time and were lower on ECX – the more liquid market – than on Nord Pool. With respect to price discovery, I demonstrate that for the first EUA futures contracts, which expired in December 2005 and 2006, both exchanges contribute to price discovery. However, for the most recent futures contracts, ECX becomes the price leader, especially in phases of high market liquidity but Nord Pool's contribution is still present from time to time.

In the last chapter[3] I motivate the importance of a properly chosen initial allocation rule for a successful implementation of a CO_2 ETS with the focus of introducing auctions. I consider a rule to be successful if it is able to (a) generate early and reliable price signals, (b) allocate the allowances to firms which need them most, and (c) to promote straightforward

---
[2]This chapter is based on the paper by Benz and Klar (2008).
[3]This chapter is based on the paper by Benz and Ehrhart (2008).

bidding, that is, firms' participation is as easy as possible and firms only have to take care of their own abatement costs. The objective of the paper is on the one hand to investigate the bidding behavior of firms when buying and selling EUAs in the auction and trading process. On the other hand, based on the behavioral results, the paper searches for an initial allocation rule that meets best the three above mentioned criteria. In this chapter, I compare several relevant allocation rules in a theoretical and experimental environment. Both environments display four policy relevant initial allocation rules in combination with emissions trading to give recommendations for a viable initial allocation mechanism for Phase 3 and beyond. This work is in reference to the recent proposal for a new Directive by the European Commission for Phase 3, that requires auctioning all allowances to the energy sector and to start with an initial auctioning share of 20% in 2013 which will increase to 100% by 2020 for the industry sector. I am able to show that the initial allocation rule that is most likely to be considered by the EU countries for Phase 3 and other international ET schemes (i.e. gratis allocation in combination with a one-sided uniform auction) does not meet the proposed functions. Instead I show that, depending on the amount of auctioned allowances, an exclusive uniform auction – in case of exclusive auctioning – or a uniform double auction – in case of partial auctioning – might be attractive candidates for Phase 3.

The next three chapters each present one idea as a self-contained unit.

Chapter 1

Modeling the Price Dynamics of CO_2 Emission Allowances

In this chapter we analyze the short-term spot price behavior of carbon dioxide (CO_2) emission allowances of the new EU-wide CO_2 emissions trading system (EU ETS). After reviewing the stylized facts of this new class of assets we investigate several approaches for modeling the returns of emission allowances. Due to different phases of price and volatility behavior in the returns, we suggest the use of Markov switching and AR-GARCH models for stochastic modeling. We examine the approaches by conducting an in-sample and out-of-sample forecasting analysis and by comparing the results to alternative approaches. Our findings strongly support the adequacy of the models capturing characteristics like skewness, excess kurtosis and in particular different phases of volatility behavior in the returns.

1.1 Introduction

By forcing the participating companies to hold an adequate stock of allowances that corresponds to their CO_2 output, the carbon market provides new business development opportunities for market intermediaries and service providers. Risk management consultants, brokers and traders buy and sell emission allowances and their derivatives. Especially for these groups, the price behavior and dynamics of this new asset class - CO_2 emission allowances - is of major importance. According to the IETA (2005) and PointCarbon (2005) previous carbon trading activities have been mostly conducted by OTC activities and brokers.

Since allowance trading has primarily been applied in the US, the majority of publications about price behavior of tradable emission allowances assesses the market for SO_2 emissions under the Acid Rain Program of the US Environmental Protection Agency (EPA). By using industrial organization models they account for changes in parameters of technology (Rezek, 1999) and electricity demand (Schennach, 2000) and their impact on the optimal equilibrium price path for SO_2 permits. There is also a number of empirical investigations on ex-post market price analysis, among them Ellerman and Montero (1998), Burtraw (1996) and Carlson, Burtraw, Cropper and Palmer (2000). For CO_2 market price simulation studies with respect to changes in market design parameters see e.g. Burtraw and Paul (2002), Böhringer and Lange (2005), Kosobud, Stokes, Tallarico and Scott (2005) or Schleich, Ehrhart, Hoppe and Seifert (2006). Kosobud et al. (2005) analyze monthly returns of SO_2 allowances with respect to other financial assets and find no statistically significant correlation between spot prices in the US and returns from various financial investments.

However, literature examining the CO_2 allowance prices from an econometric or risk management angle is rather sparse. Exceptions include Daskalakis, Psychoyios and Markellos (2006); Paolella and Taschini (2006); Seifert, Uhrig-Homburg and Wagner (2008) and Uhrig-Homburg and Wagner (2006). While Uhrig-Homburg and Wagner (2006) investigate the success chances and optimal design of derivatives on emission allowances, Seifert et al. (2008) develop a stochastic equilibrium model reflecting in a stylized way the most important features of the EU ETS and analyze the resulting CO_2 spot price dynamics. Their main findings are that an adequate CO_2 process does not necessarily have to follow any seasonal patterns. It should possess the martingale property and exhibit a time- and price-dependent volatility structure. Paolella and Taschini (2006) provide an econometric analysis addressing the unconditional tail behavior and the heteroskedastic dynamics in the returns on CO_2 and SO_2 allowances. They find that models based on the analysis of fundamentals or on the future-spot parity of CO_2 yield implausible results due to the complexity of the market and advocate the use of a new GARCH-type structure. Finally, examining emission allowance prices and derivatives, Daskalakis et al. (2006) find some

1.1. INTRODUCTION

evidence that market participants adopt standard no-arbitrage pricing.

We differ from the analysis of the mentioned papers by also concentrating on the out-of-sample performance of the models with respect to forecasting. In particular, we evaluate price, volatility and density forecasts for the different approaches what can be considered as a substantial issue in managing price risk. With an increasing range of new instruments (e.g. spot, forwards, futures, etc.) the carbon market is steadily gaining in complexity. Risk managers and traders constantly have to hedge their positions against irregular and unexpected carbon price fluctuation. Hence, they are not only interested in the long-term perspective of emission allowance prices but also in short-term price dynamics of the assets. Having a reliable pricing and forecast model will allow companies, investors and traders to realize efficient trading strategies, risk management and investment decisions in the carbon market.

The aim of this chapter is to provide an analysis of the short-term spot price behavior of CO_2 emission allowances focusing on the price dynamics and changes in the volatility of the underlying stochastic price process. Since CO_2 emission allowances are a new trading good in the European commodity market, there is not much historical data available. By studying the new market mechanism and analyzing first empirical data we consider the appropriateness of several stochastic price processes. The suggested econometric models can be used in particular for short-term forecasting and Value-at-Risk (VaR) calculation. Thus, they could be especially helpful for risk managers or traders in the market, but might also enable companies to monitor the costs of CO_2 emissions in their production process. The remainder of the chapter is organized as follows. Section 1.2 provides a brief introduction into the new market mechanism for CO_2 emission allowances and a classification of this new commodity. Section 1.3 presents stochastic approaches for modeling the price dynamics of CO_2 allowances, namely regime-switching and AR-GARCH models. Section 1.4 provides results from the empirical analysis of CO_2 allowance prices and returns. In an in-sample and out-of-sample analysis we benchmark the models against other approaches, including autoregressive processes and a simple i.i.d. Gaussian model. Section 1.5 concludes and gives suggestions for future work.

1.2 Market Mechanism and Instruments

1.2.1 The EU ETS and Classification of Emission Allowances

Generally, a company's stock of emission allowances determines the degree of allowed plant utilization. Thus, a lack of allowances requires from the company either some plant-specific or process improvements, a cut- or shutdown of the emission producing plant or the purchase of additional allowances and emission credits. With the latter two alternatives CO_2 becomes a new member of the European commodity trading market. There is, however, a fundamental difference between trading in CO_2 and more traditional commodities. What is actually sold is a lack or absence of the gas in question. Sellers are expected to produce fewer emissions than they are allowed to, so they may sell the unused allowances to someone who emits more than her allocated amount. Therefore, the emissions become either an asset or a liability for the obligation to deliver allowances to cover those emissions (PointCarbon, 2004).

Benz and Trück (2006) point out the differences between emission allowances and classical stocks. While the demand and the value of a stock is based on profit expectations of the underlying firm, the CO_2 allowance price is determined directly by the expected market scarcity induced by the current demand and supply at the carbon market. Notably, firms by themselves are able to control market scarcity and hence the market price by their CO_2 abatement decisions. It is important to note that the annual quantity of allocated emission allowances is limited and already specified by the EU-Directive for all trading periods. Additionally, in case of an intertemporal ban in banking of CO_2 emission allowances, the certificates have a limited duration of validity. The value of an individual allowance expires after each commitment period. Allowing for an intertemporal transfer, the allowances only lose their value once used for covering CO_2 emissions.

A more appropriate approach in specifying CO_2 emission allowances is their consideration as a factor of production (Fichtner, 2005). The shortage of emission allowances by reducing the emissions cap for the commitment periods classifies the assets as 'normal' factors of production. They can be 'exhausted' for the production of CO_2 and after their redemption or at the end of the commitment period when they expire, they are removed from the market. Additionally, if there is an intertemporal ban on banking between the commitment periods– as it was the case from the pilot phase to the CP I – all allowances become worthless at the end of the periods and thus are non-storable. On the other side, if banking is allowed the validity of allowances is renewed for the upcoming commitment period. Accordingly, it seems more adequate to compare the right to emit CO_2 with other operating materials or commodities than with a traditional equity share and hence to adopt rather commodity than stock pricing models (see Section 1.3).

1.2.2 Price Determinants of CO_2 Emission Allowances

Having gained knowledge about the particularities of the new assets, it is essential for carbon market players to learn about their price dynamics in order to realize trading strategies, risk strategies and investment decisions. In this section, we identify the key price determinants of the CO_2 emission allowances, which an appropriate commodity pricing model should be able to display. According to the investigation of SO_2 permit prices by Burtraw (1996), we categorize the principle driving factors of CO_2 allowance prices into (i) policy and regulatory issues and (ii) market fundamentals that directly concern the production of CO_2 and thus demand and supply of CO_2 allowances.

Monitoring price sources from part (i), it is reasonable to assume that they have a long-term impact on prices. However, for our model we are only interested in those policy issues, which additionally have a rather low probability for an exact forecast. Changes in policy directives or regulations may have substantial consequences on actual demand and supply and thus on short-term price behavior of emission allowances. This is comparable with the effect that some good or bad news published on an individual company may have on its share price. In the carbon market these could be decisions and announcements concerning the National Allocation Plans (NAPs) that set the rules and reduction targets (e.g. NAP revisions or cut of national emission caps). Hence, the consequences of changes in such regulatory or policy issues may be sudden price jumps, spikes or phases of extreme volatility in allowance prices.

Note that aspects concerning the regulatory framework like explicit trading rules (e.g. intertemporal trading), the linkage of the EU ETS with the market of project-based mechanisms and/or with the Kyoto Market in the future have an important impact on prices, too. However, they are the result of a long discussion process whose consequences have to be studied extensively in advance, see e.g. Anger (2008); Schleich et al. (2006) and Seifert et al. (2008). Hence, market participants might be able to hedge themselves against these foreseen 'price risks' in the long term. They are not incorporated in our econometric models focusing on short-term price behavior.

Incorporating part (ii), allowance prices may also show phases of specific price behavior due to fluctuations in production levels. In general, CO_2 production depends on a number of factors, such as weather data (temperature, rain fall and wind speed), fuel prices and economic growth. Especially unexpected (environmental) events[1] and changes in fuel spreads will shock the demand and supply side of CO_2 allowances and consequently market prices. Cold weather increases energy consumption and hence CO_2 emissions through power and heat generation; rainfall and wind speed affect the share of non-CO_2 power generating sources and thus emission levels. A short term measure for the power and heat sector to invest in CO_2 abatement projects are the relative costs of coal and cleaner fossil fuels such as oil and natural gas. Europe's cheapest path is to switch from coal-fired to gas-fired power generations, which need less than half of the allowances required by their coal-fired counterparts to produce the same amount of electricity.[2] Therefore, this source of price uncertainty may have a rather short or medium-term impact on market liquidity of the allowances that possibly increases volatility of the allowance prices.

Overall, we assume that allowance prices and returns will exhibit different periods of price behavior including price jumps or spikes as well as phases of high volatility and

---

[1] E.g. power plant breakdowns (nuclear-, coal-fired- or hydroelectric power plants) where more emission intensive power stations have to be set up or unexpected environmental disasters (forest fire, earthquakes, etc.) shock the demand and supply side of CO_2 allowances.

[2] Depending on the capacity, the turning-on of gas turbines only takes several minutes (BMWT, 2006).

heteroscedasticity in returns. It is the challenge of an appropriate stochastic model to capture such a price pattern.

1.3 Modeling the Price Dynamics of CO_2 Emission Allowances

In this section we incorporate the aforementioned characteristics of CO_2 allowances and their price determinants, in particular the different phases of volatility behavior and the dependence of the variability of the time series on its own past in an adequate stochastic model. Hence, we suggest models allowing for heteroscedasticity like ARCH, GARCH or regime-switching models. While the former two suggest a unique stochastic process but conditional variance, the latter divides the observed stochastic behavior of a time series into several separate phases with different underlying stochastic processes.

1.3.1 GARCH Models

While the traditional linear ARMA-type models assume homoscedasticity, i.e. a constant variance and covariance function, the autoregressive conditional heteroskedastic (ARCH(p)) time series model of Engle (1982) was the first formal model which successfully addressed the problem of heteroskedasticity. In this model the conditional variance of the time series $(y_t)_{t \geq 0}$ is represented by an autoregressive process (AR), namely a weighted sum of squared preceding observations:

$$y_t = \varepsilon_t \sigma_t, \quad \text{with} \quad \sigma_t^2 = a_0 + \sum_{i=1}^{q} a_i y_{t-i}^2, \tag{1.1}$$

where ε_t are i.i.d. with zero mean and finite variance (typically it is assumed that $\varepsilon_t \overset{iid}{\sim}$ N(0,1)).

In practical applications to financial time series data it turns out that the order q of the calibrated model is rather large (Pagan, 1996). However, if we let the conditional variance depend not only on the past values of the time series but also on a moving average of past conditional variances the resulting model allows for a more parsimonious representation of the data. This model, the generalized autoregressive conditional heteroskedastic model (GARCH(p,q)) put forward by Bollerslev (1986) and Taylor (1986) is defined as

$$y_t = \varepsilon_t \sigma_t, \quad \text{with} \quad \sigma_t^2 = \alpha_0 + \sum_{i=1}^{q} \alpha_i y_{t-i}^2 + \sum_{j=1}^{p} \beta_j \sigma_{t-j}^2, \qquad (1.2)$$

where ε_t are as before. The coefficients have to satisfy $\sum \alpha_i + \sum \beta_j < 1$ and $\alpha_i, \beta_j \geq 0$, $\alpha_0 > 0$ to ensure stationarity and a conditional variance that is strictly positive. Identification and estimation of GARCH models is performed by maximum likelihood estimation, e.g. documented by Brooks, Burke and G. (2001).

Obviously, the GARCH model is especially designed to model the conditional volatility of a time series. However, the variance equation can be coupled for example with an AR(r) process for the mean of the time series

$$y_t = c + \sum_{k=1}^{r} \phi_k y_{t-1} + \varepsilon_t, \qquad (1.3)$$

where $\phi_k < 1$ and c denote real constants. Then the model provides a promising approach to model both the mean and the variance of the considered time series – the AR-GARCH model. The literature on GARCH or AR-GARCH models for analyzing financial time series is extensive. Applications to models for commodities include Garcia, Contreras, van Akkeren and Garcia (2005); Morana (2001); Mugele, Rachev and Trück (2005); Ramirez and Fadiga (2003).

1.3.2 Regime-Switching Models

The second class of pricing models that we suggest are the so-called regime-switching models. Hereby, we follow the idea of Goldfeld and Quandt (1973); Hamilton (1989, 1990) who introduced regime-switching models and successfully suggested their use for financial time series. There are also a number of recent publications where the models are used to describe asset returns in financial markets (Kanas, 2003; Kim and Nelson, 1999; Kim,

1.3. MODELING THE PRICE DYNAMICS OF CO_2 EMISSION ALLOWANCES

Piger and Startz, 2004; Schaller and van Norden, 1997). In the last decade the models also became especially popular for modeling electricity spot prices (Bierbrauer, Trück and Weron, 2004; Ethier and Mount, 1998; Huisman and Mahieu, 2001; Weron, Bierbrauer and Trück, 2004; Haldrup and Nielsen, 2004). Due to their promising features of modeling different regimes of price and volatility behavior we suggest the approach also for modeling CO_2 emission allowances' logreturns.

In general, regime-switching models divide the time series into several phases that are called regimes. For each regime one can define separate and independent underlying price processes. The literature distinguishes between two main classes of regime-switching models (Franses and van Dijk, 2000). In the first one, the regime can be determined by an observable variable. Consequently, the regimes that have occurred in the past and present are known with certainty. In the second class the regime is determined by an unobservable, latent variable. In this case we can never be certain that a particular regime has occurred at a particular point in time, but we can only assign or estimate probabilities of their occurrences. In the following we will suggest to use the second class of models that is often referred to as Markov regime-switching models. We argue that it is rather questionable to assume that the regime-switching mechanism is simply governed by a fundamental variable or the price process itself. As described in Section 1.2.2, spot prices or returns of CO_2 emission allowances are the outcome of a vast number of variables including fundamentals (like weather or macroeconomic variables) but also the unquantifiable regulatory, policy and sociological factors that can cause an unexpected and irrational buyout or lead to price jumps and periods of extreme volatility.

Hence we assume that the switching mechanism between the states is governed by an unobserved random variable R_t. For example, a model with two regimes follows a Markov chain with two possible states, $R_t = \{1, 2\}$. Hereby, the spot price or return may be assumed to display either low or very high volatility at each point in time t, depending on the regime $R_t = 1$ or $R_t = 2$. Consequently, we have a probability law that governs the transition from one state to another, while the processes y_{t,R_t} for each of the two regimes are supposed to be independent from each other. Further, a transition matrix $\mathbf{Q}$ contains the probabilities q_{ij} of switching from regime i at time t to regime j at time $t+1$, for $i,j = \{1,2\}$:

$$\mathbf{Q} = (q_{ij}) = \begin{pmatrix} q_{11} & q_{12} \\ q_{21} & q_{22} \end{pmatrix} = \begin{pmatrix} q_{11} & 1-q_{11} \\ 1-q_{22} & q_{22} \end{pmatrix}. \quad (1.4)$$

Due to a property of Markov chains the current state R_t only depends on the past through the most recent value R_{t-1}:

$$P\{R_t = j | R_{t-1} = i, R_{t-2} = k, \ldots\} = P\{R_t = j | R_{t-1} = i\} = p_{ij} \quad (1.5)$$

Consequently the probability of being in state j at time $t+m$ starting from state i at time t is given by

$$(P(R_{t+m} = j \mid R_t = i))_{i,j=1,2} = (\mathbf{Q}')^m \cdot e_i, \quad (1.6)$$

where $\mathbf{Q}'$ denotes the transpose of $\mathbf{Q}$ and e_i denotes the ith column of the 2×2 identity matrix. The variation of regime-switching models is due to both the possibility of choosing the number of regimes and different stochastic processes assigned to each regime. In the literature, often a mean-reverting process with Gaussian innovations is used for the various regimes (Bierbrauer et al., 2004; Huisman and Mahieu, 2001) while other model specifications are possible and straightforward. Hamilton (1989) for example suggests an autoregressive process of higher order for both regimes, while for return modeling a white noise process for either regime may be adequate (Kim and Nelson, 1999; Schaller and van Norden, 1997).

Given the stated assumptions about the price behavior of CO_2 emission allowances, applying regime-switching models may be a promising approach. It reflects the concept of having a systematic change between stable and unstable states which results from fluctuations in demand and supply on markets as assumed for the CO_2 allowance market in the previous section. Furthermore, the model allows for several consecutive price jumps or extreme returns that are important when talking about risk management and pricing of derivative instruments.

Unfortunately, parameter estimation of the two underlying processes is not straightforward since the regime is latent and hence not directly observable. Hamilton (1990) introduced an application of the EM algorithm by Dempster, Laird and Rubin (1977) for the estimation procedure. The regime R_t is modeled as the outcome of an unobserved two-state Markov chain with $R_t = \{1,2\}$. Additionally, the estimation process needs a stochastic process for each regime y_{t,R_t}, $R_t = \{1,2\}$, $t = 1, \ldots, T$ and a transition matrix $\mathbf{Q}$. The EM algorithm uses an iterative procedure to collect and estimate the parameter set θ based on an initial parameter estimate $\hat{\theta}^{(0)}$. Then each iteration of the EM algorithm generates new estimates $\hat{\theta}^{(n+1)}$ of the unknown parameter set based on the previously calculated vector set $\hat{\theta}^{(n)}$. The algorithm stops as soon as the change in the loglikelihood function (LLF) is small enough, i.e. when the process has converged. Hamilton (1990) shows, that each iteration cycle of the sample increases the LLF and the limit of this sequence of estimates reaches a (local) maximum of the LLF. For a detailed technical specification refer to Dempster et al. (1977) and Hamilton (1994).

1.4 Empirical Results

1.4.1 The Data

In this section we investigate the appropriateness of the suggested time series models for logreturns of daily EU CO_2 allowance (EUA) prices. The considered time period is from January 3, 2005 - December 29, 2006. Hereby, the data from period January 3, 2005 - December 30, 2005 is used for the calibration of the models, while the period January 3, 2006 - December 29, 2006 is used for out-of-sample testing.

Figure 1.1 shows a plot of daily EUA prices for the period August 27, 2003 - December 29, 2006. The data are provided by Spectron one of the major brokers in the energy trading industry and stems from OTC transactions (Spectron, 2006).

The operational trade with EUAs already began in 2003, before the official agreement on the EU ETS. In the "pre-2005" period, the traded volume was quite low, at some days even zero as the highest bidder price was smaller than the lowest seller price. The market

Figure 1.1: Daily EUA Prices from August 27, 2003 - December 29, 2006

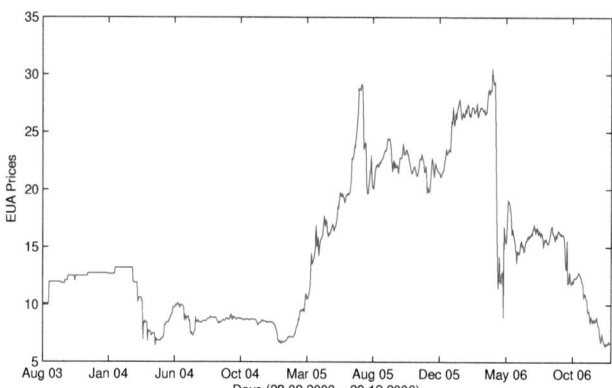

price then was just determined by the mean of the two figures. Bid-ask spreads were quite large, often exceeding 4 Euro, indicating that prices could not be considered as stemming from real trading activities. One should note, that prices before 2005 are forward prices on a not yet traded underlying. Hence they have to be considered with care when they are compared to spot prices starting 2005. Ulreich (2005) points out that the pre-2005 period was more useful for setting up the infrastructure for the official start of the EU ETS in 2005 than getting important market price signals.

Having a closer look at the run of the curve in Figure 1.1, our model for EUAs' key price drivers from Section 1.2.2 can be verified.[3] Before the EU-parliament agreed on the introduction of the EU ETS in July 2003 and before the first suggestions for National Allocation Plans (NAPs) were published at the end of 2003, prices were quite stable.[4] Both announcements led to an increase in prices. Because of the initially generous allocation of

---

[3]Price data have not been available for July 2003. Consequently, the price path in Figure 1.1 starts in August 2003.

[4]For each commitment period, the member state have to develop a NAP that sets the reduction targets for the covered sectors and how it is divided among the covered installations.

1.4. EMPIRICAL RESULTS

allowances to the countries prices calmed down again between February and March 2004. Reviewing and accepting the NAPs in the second half of the year, prices increased and settled down around 9 Euro. As the main framework of the trading scheme has been defined, the price determinants became more fundamental after January 2005 (Ulreich, 2005).

The market began to respond increasingly to changes in the underlying energy markets and the weather. We find that prices initially fell due to a quite mild climate and high supply of wind energy from Scandinavia and North Germany. At the end of January an extreme cold snap and constant high UK gas and oil prices, compared to relatively low coal prices, led to a drastically price increase (PointCarbon, 2005). This effect was boosted by an extremely dry summer in July 2005 in the southwest of Europe. The absence of necessary rainfall prevented full utilization of hydraulic plants, especially in Spain. Additionally, the lack of cooling water for nuclear plants led to higher emission-intensive power plant utilization and therefore increased the demand for CO_2 permits. By mid of July 2005 prices peaked at 29.15 Euro. Since then, during the last four months of 2005 prices fell and stabilized around 22 Euro. However, in the beginning of 2006, a renewed increase in the price level could be observed to approximately 27 Euros by the end of March. Reasons for that, once again, may be the extremely long and cold winter in 2005/2006.

May 2006 saw the completion of the first full compliance cycle of the EU ETS with the publication of the 2005 verified emissions data. But already in April 2006, it became clear that corporate participants had been granted around 10% more allowances than they actual needed to cover their 2005 emissions. Consequently, surplus EUAs flooded the market, prices crashed 60% within one week, from a high of around 30 Euros per ton of CO_2 to 11 Euros. Traders began to express the fear that the emissions price would drop to zero. With so many allowances being given out, even factors such as the fluctuations in the use of fossil fuel associated with yearly variations in weather are now playing havoc with demand, putting prices in doubt. Then, prices stayed volatile, especially since no European government wanted to be the first to reduce radically the number of allowances granted to the industry. In June and July 2006 the EUA market recovered as industrial companies

Figure 1.2: Daily EUA logreturns from January 3, 2005 - December 29, 2006

The figure displays daily EUA logreturns for both calibration and test period from January 3, 2005 - December 29, 2006.

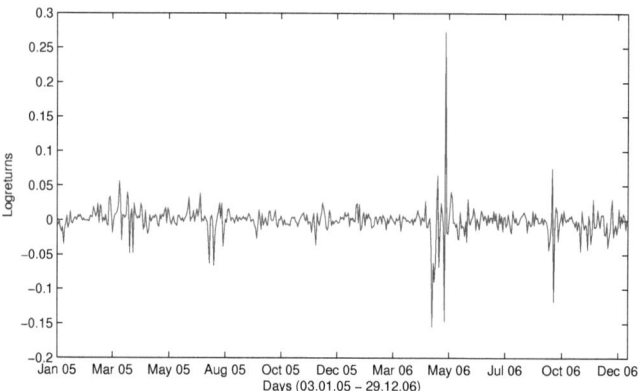

started selling EUAs to utilities and financial players and the hot, dry July in Europe led to higher demand for electricity even as hydro resources were low and nuclear resources were off-line pushing the spot price of EUAs higher to around 16 Euros. In September, EUA spot prices declined sharply following the collapse of spot prices of natural gas in Europe. Over the next months, the EU Commission began to review the proposed allocation plans by Member States for Phase 2 of the EU ETS (NAPs II).

Overall, we can conclude that spot price behavior in the CO_2 emission allowance market confirm our motivation for looking at a price model that deals with volatile price processes induced by short-term factors like the spread between fuel prices, precipitation, summer and winter temperature and the setup of a trading environment.

1.4.2 Analyzing Logreturns of CO_2 Allowances

Figure 1.2 shows a plot of the EUA logreturns $y_t = log(S_t) - log(S_{t-1})$ for the whole considered period while a summary statistics for the EUA prices S_t and logreturns y_t in the calibration and forecasting period is presented in Table 1.1.

Table 1.1: Summary statistics for the EUA logreturns

The table reports the summary statistics for the EUA logreturns y_t for the in-sample period January 3, 2005 - December 30, 2005 and out-of-sample period January 3, 2006 - December 29, 2006.

Series	N	Mean	Median	Min	Max	Std Dev	Skew	Kurt
In-Sample	256	0.0037	0.0046	-0.1528	0.1298	0.0319	-0.83	8.57
Out-of-Sample	253	-0.0047	-0.0017	-0.3551	0.6267	0.0653	2.06	43.13

Obviously, the data show heteroskedasticity and volatility clustering. In 2005, the calibration period, in March as well as during the very dry summer in July the logreturns exhibit a clearly increased volatility. As a consequence, both maximum positive and negative logreturns could be observed during this period. The former was 0.1298 Euro on July 4, 2005 while the latter could be observed on July 14, 2005 and was -0.1528 Euro. For the logreturns we get a skewness parameter of $s = -0.83$ and a kurtosis of $k = 8.57$ in the calibration period and $s = 2.06$ and $k = 43.13$ in the out-of-sample period. We conclude that in both periods the logreturns exhibit skewness and excess kurtosis. However, logreturns are left-skewed during the calibration period and right-skewed during the out-of-sample period. Figure 1.3 provides the empirical distribution of the logreturns for the whole period from January 3, 2005 - December 29, 3006, including a fit of the normal distribution to the data. Due to asymmetry, excess kurtosis and heavy tails, the normal distribution doesn't fit the data very well. Hence, alternative models allowing for changes in the volatility structure, asymmetry and excess kurtosis should provide a better fit to the time series.

Figure 1.3: Empirical distribution and Gaussian fit to EUA logreturns

The figure displays the empirical distribution obtained by kernel estimator (solid) and Gaussian (dashed) fit to daily EUA logreturns from January 3, 2005 - December 29, 2006.

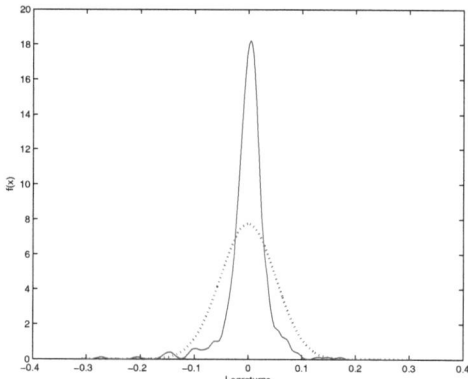

1.4.3 Time series Models

After examining daily logreturns of EUA, in a second step we investigate the adequacy of the suggested AR-GARCH and regime-switching models for the time series. To benchmark our estimation results, we also compare them to the results of a simple normal distribution for the logreturns as well as to an AR(r). For the AR-GARCH models, we have to specify both the mean and variance equation. For the mean equation we chose the same AR(r) process as in Equation (1.3). However, taking non-constant variance in the residuals into account, the noise terms are not just i.i.d. $(0,\sigma^2)$ but are given by a GARCH(p, q) process. All models are estimated by using maximum likelihood estimation.

For analyzing returns of financial time series with regime-switching models, Kim and Nelson (1999) suggest a white noise processes for both regimes, while Schaller and van Norden (1997) investigate whether stock market logreturns are drawn from Gaussian distributions with the same or different means and variances. The structure and parameter estimation of such models have been described in the previous section. Hence, the remaining task

1.4. EMPIRICAL RESULTS

consists of specifying the two stochastic processes $y_{t,1}$ and $y_{t,2}$. Following the literature we suggest either a white-noise process with different mean and variance for both regimes, $R_t = \{1,2\}$ (Kim and Nelson, 1999; Schaller and van Norden, 1997) or a mean-reverting process for the 'base regime' ($R_t = 1$) (Ethier and Mount, 1998; Bierbrauer et al., 2004) while the 'spike regime' ($R_t = 2$) is modeled by independent and identically distributed realizations of a Gaussian distribution (Huisman and Mahieu, 2001). In summary, we are considering the following stochastic processes for the regimes:

For the base regime we either use

$$y_{t,1} \stackrel{iid}{\sim} N(\mu_1, \sigma_1^2), \quad t \in \mathbb{N} \tag{1.7}$$

in the model labeled 'Gaussian' or

$$y_{t,1} = \phi y_{t-1,1} + c + \varepsilon_t, \quad t \in \mathbb{N} \tag{1.8}$$

in the model labeled 'Mean Reversion' (MR), while for the spike regime we suggest

$$y_{t,2} \stackrel{iid}{\sim} N(\mu_2, \sigma_2^2), \quad t \in \mathbb{N}. \tag{1.9}$$

for both model specifications. The innovations ε_t in Equation (1.8) are assumed to be i.i.d. centered normal ($\varepsilon_t \stackrel{iid}{\sim} N(0, \sigma^2)$), $\phi < 1$, c denote real constants and N displays the normal distribution with parameters μ_i and σ_i^2, $i = \{1,2\}$. Process (1.8) is the discrete version of a standard Vasiček model. We choose to specify the stochastic processes in discrete time, simplifying the estimation procedure which is based on a discrete-time sample.

1.4.4 In-Sample Results

In the following we discuss the estimation and in-sample results for the suggested models. We first consider the results from fitting a Gaussian distribution and an AR(r) process to the logreturns. For the simplest model – fitting a normal distribution to the data –

we obtain the parameters $\mu = 0.0037$ and $\sigma = 0.0319$. As indicated by Schwartz (1997), many commodity prices are in general regarded to be mean reverting. In discretized form, a mean-reverting process is then equivalent to a Gaussian AR(1) process:

$$r_t = c + \phi r_{t-1} + \varepsilon_t, \qquad (1.10)$$

where $\phi < 1$ and c denote real constants and the innovations ε_t are assumed to be i.i.d. normal centered ($\varepsilon_t \stackrel{iid}{\sim} N(0, \sigma^2)$). The parameter estimates for the AR(1) process are $c = 0.0033$, $\phi = 0.2122$ and $\sigma_\varepsilon = 0.0313$. Note that also AR processes of higher order were tested, but according to the Schwarz Bayesian Criterion the AR(1) specification was considered as optimal. Obviously, both model specifications provide almost the same estimate for the variance. Further, the added explanatory power by the AR(1) process is rather limited what is also indicated by the only slightly increasing LLF in Table 1.4.

The residuals obtained by the fitted AR(1) process seemed to exhibit non-constant variance. Testing with the Lagrange multiplier ARCH test statistics (Engle, 1982) the heteroskedastic effects are highly significant. To capture this behavior also a GARCH(p, q) model was calibrated to the data. Hereby, for the mean equation we chose an AR(1) process, while for the variance equation we test different GARCH specifications. It turns out that parameter estimates for higher orders of p or q are not significant. Thus, we obtain the simple setup of an AR(1)-GARCH(1,1) model and the following variance equation:

$$\varepsilon_t = u_t \sigma_t, \quad \text{with} \quad \sigma_t^2 = k + \alpha \varepsilon_{t-1}^2 + \beta \sigma_{t-1}^2, \qquad (1.11)$$

where u_t is i.i.d. with zero mean and finite variance and k, α, β are real constants.

Note that to simplify the notation, in the following we will refer to the AR-GARCH model as GARCH model. Estimation results of the GARCH model are provided in Table 1.2, all estimated coefficients of the model are significant. Figure 1.4 displays graphs of the innovation, conditional standard deviation and the logreturn time series for the estimated model. As expected, during times of extreme positive or negative returns, the estimated

1.4. EMPIRICAL RESULTS

Figure 1.4: innovations, conditional standard deviations and logreturns of AR(1)-GARCH(1,1)

The figure displays innovations, conditional standard deviations and logreturns of the estimated AR(1)-GARCH(1,1) model for the in-sample period January 3, 2005 - December 30, 2005.

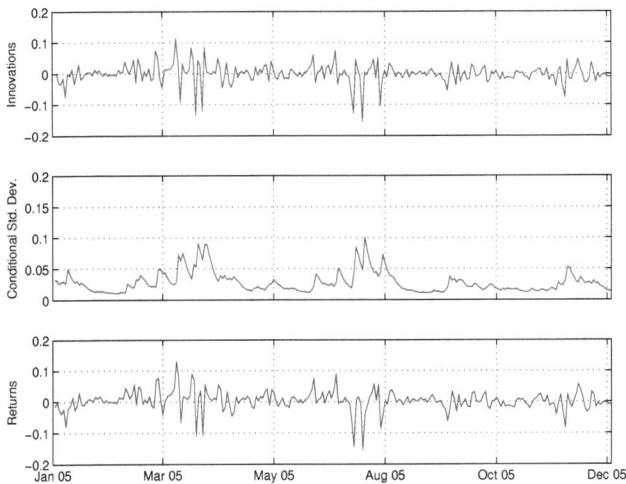

conditional variance increases substantially. While the estimates of the conditional standard deviation are clearly below 0.05 during rather quiet periods, they increase up to 0.1 during volatile periods like in March/April or July/August 2005. Obviously, the GARCH model describes the data better than a simple normal distribution or an AR process and seems more appropriate for the price dynamics of EUA logreturns.

Finally, we estimate the two different regime-switching specifications for the in-sample period. Parameter estimates are displayed in Table 1.3. We first compare the estimated standard deviation of the 'Gaussian' model – a mixture of two normal distributions – to the simple model of a single normal distribution for the logreturns. We find that the standard deviation of the 'naive' fit lies between the two estimated standard deviations $\sigma_1 = 0.0122$ for the base regime and $\sigma_2 = 0.0476$ for the spike regime. The same is true for μ being

Table 1.2: Parameter estimates of the AR(1)-GARCH(1,1) model

This table reports the parameter estimates of the AR(1)-GARCH(1,1) model for the in-sample period January 3, 2005 - December 30, 2005.

	Coefficient	Std. Error	t-Statistic
	Mean Equation		
c	0.00243	0.00129	1.881
ϕ	0.29713	0.06652	4.467
	Variance Equation		
k	3.7309e-005	1.0445e-005	3.5720
α	0.53253	0.036733	14.4973
β	0.36747	0.055515	6.6193

higher than the expected logreturn in the base regime $\mu_1 = 0.0029$ but clearly lower than $\mu_2 = 0.0050$ in the spike regime. In terms of the variance of the regimes, very similar results are obtained also for the model 'MR'. However, here the expected value with $\mu_1 = 0.0040$ for the base regime and $\mu_2 = 0.0042$ for the spike regime do not differ that much. Overall, both regime-switching models seem to distinguish between two phases of logreturns: one phase with clearly higher variance for the volatile periods, and one for the less volatile period yielding a lower mean and variance in the returns.

Moreover, the estimated volatility in the two regimes is of special interest, since in the empirical data we observe periods of very low volatility being followed by phases of much higher volatility. In both specifications the estimates for σ_2 are approximately four times higher than σ_1. This results in a variance about 16 times higher for the spike regime than for the base regime. In both models the probability of being in the base regime is higher, approximately 58% for the model 'Gaussian' and 66% for the model 'MR', while the spike regime has the probability of approximately 42% and 34%. Consequently, the probability for remaining in the same regime p_{ii} is higher for the base regime: we have approximately $p_{11} = 88\%$ for both model specifications. This indicates that a change in the regimes does not occur frequently. We conclude, that for the considered data the estimated parameters

1.4. EMPIRICAL RESULTS

Table 1.3: Estimation results with the two-state regime-switching model

Panel (a): Estimation results for logreturns with the two-state regime-switching model with a simple normal distribution in both regimes (model 'Gaussian').
Panel (b): Estimation results for logreturns with a two-state regime-switching model with a mean-reversion process in the base regime and a Gaussian distribution in the spike regime (model 'MR').

Panel (a): Model 'Gaussian'

Regime	Parameter Estimates			Statistics	
	μ_i	σ_i	p_{ii}	$P(R_t = i)$	$E(y_{t,i})$
base ($i=1$)	0.0029	0.0122	0.8768	0.5814	0.0029
spike ($i=2$)	0.0050	0.0476	0.8289	0.4186	0.0050

Panel (b): Model 'MR'

Regime	Parameter Estimates					Statistics	
	ϕ	c	μ_i	σ_i	p_{ii}	$P(R_t = i)$	$E(y_{t,i})$
base ($i=1$)	0.2661	0.0029	-	0.0137	0.8834	0.6598	0.0040
spike ($i=2$)	-	-	0.0042	0.0513	0.7738	0.3402	0.0042

Figure 1.5: Logreturns of EUA prices and probability of being in the spike regime

The figure displays logreturns of EUA prices from January 3, 2005 to December 30, 2005 together with the probability of being in the spike regime
Top panel: Logreturns of EUA prices from January 3, 2005 to December 30, 2005.
Bottom panel: Probability of being in the spike regime for the defined two-regimes 'Gaussian' model for the same period.

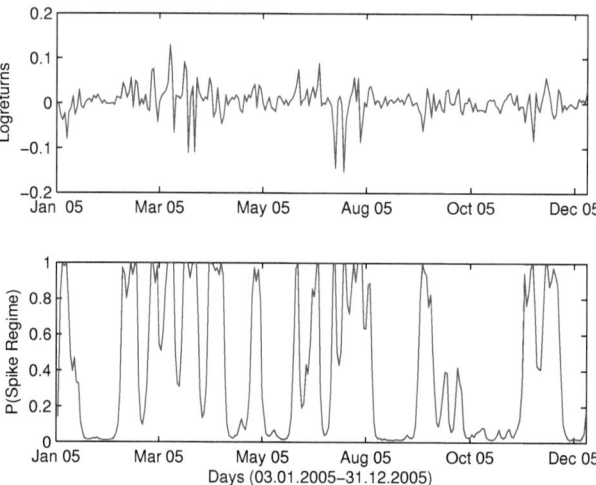

are meaningful and can be interpreted in terms of an adequate distinction between the different phases of volatility behavior. Furthermore, we find that both regime-switching specifications lead to similar results.

Another decisive question is whether the models are able to significantly distinguish between the regimes in terms of the assigned probabilities to either one of the two regimes. Therefore, Figure 1.5 provides a graphs that shows the original logreturn series and the corresponding estimated probability of being in the spike regime for the model 'Gaussian'. Note, as the results for the model 'MR' are quite similar we forbear from showing these graphs. Obviously, in times of extreme volatility behavior of the logreturns we observe

1.4. EMPIRICAL RESULTS

analogously a clear distinction between the two regimes. Most notably, for the periods March/April as well for June to August and November/December the model assigns many observations to the spike regime. This behavior can be explained very well by means of the two principle driving factors introduced in Section 1.2.2. First, at beginning of March the cold weather forced Spanish and French firms to enter the market and thus increased the demand side. Additionally, the sudden increase in spot prices of UK natural gas and oil in March and again at the beginning of April increased the spot price for CO_2. Besides, within the two months important decisions were made with respect to NAPs: In the mid of March is was announced to cut the Polish and Czech NAP and at the same time to possibly increase allocation for UK and Italy. The decision of the European Council was published at beginning of April to reduce emissions by 15-30% until 2020 and by 60-80% by 2050. However, at the end of April the CO_2 spot price recovered drastically and move to the base regime due to a constantly falling oil prices on the one side and on the other side due to the launch of organized trading platforms in May (European Climate Exchange (ECX) and PowerNext) which made speculative trading more important.

As the distinction between the two regimes is even better for the second period from June 1, 2005 to August 31, Figure 1.6 provides a closer look: in this phase prices were mainly driven by fundamentals. Most notably the dry summer period in July, which boasted emissions (especially in Spain and France). Besides, over the whole period, again high oil and gas price (relative to coal) drove the the price for CO_2. At the end of 2005 the market reacted again to three important market announcements, the World Bank's forecast for CER supply, the drastic cut of the Italian NAP by 10 mt of CO_2 and the agreement on a cap-and-trade program by seven US states, the Reginal Greenhouse Gas Initiative (RGGI). To sum up, from the probabilities in Figures 1.5 and 1.6 it becomes obvious that with high probability the model assigns most of the logreturns to either one of the two regimes. This indicates the model's ability to distinguish between the two regimes of different volatility. For model evaluation, we also examine the Akaike information criterion (AIC) and the Bayesian information criterion (BIC) for the estimated models (see Table 1.4). We find that according to the chosen parsimony model criteria our results are confirmed: the

Figure 1.6: Logreturns of EUA prices and probability of being in the spike regime

The figure displays logreturns of EUA prices from June 1, 2005 to August 31, 2005 together with the probability of being in the spike regime
Top panel: Logreturns of EUA prices from June 1, 2005 to August 31, 2005.
Bottom panel: Probability of being in the spike regime for the defined two-regime 'Gaussian' model for the same period

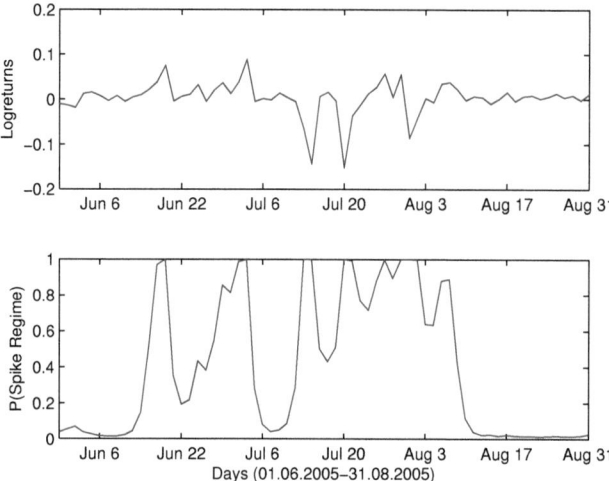

GARCH and regime-switching models clearly outperform the approach of fitting a normal distribution or an AR(1). The results for a the GARCH model and the regime-switching models 'Gaussian' and 'MR' are quite similar. While for the AIC the best results are obtained for the regime-switching model with an AR process for the base regime ('MR'), for the BIC the GARCH model gives the best results. However, similar to the results for the log-likelihood function, the differences between the two regime-switching and the GARCH model are quite small. We conclude that as far as in-sample results are concerned, GARCH and regime-switching models are adequate approaches for modeling EUA logreturns.

1.4. EMPIRICAL RESULTS

Table 1.4: Model evaluation by the log-likelihood and information criteria

The table reports the number of parameters k, log-likelihood, Akaike information criterion (AIC), and Bayesian information criterion (BIC) for the estimated models.

	k	LLF	AIC	BIC
i.i.d. Normal	2	519.45	-1034.90	-1027.82
AR(1)	3	525.78	-1045.56	-1034.94
GARCH(1,1)	5	575.72	-1141.44	-1123.73
'Gaussian'	6	574.55	-1137.10	-1115.85
'MR'	7	578.96	-1143.92	-1119.13

1.4.5 Forecasting Results

We conduct an out-of-sample analysis of the models by comparing one-day-ahead point and density forecasts for EUA logreturns for the period January 3, 2006 to December 29, 2006. Hereby, both a static approach using the estimated models for the whole out-of-sample period as well as a recursive and rolling window technique with reestimation of the parameters after each day were examined. For the recursive window approach the initial estimation date is fixed and additional observations are added one at a time to the estimation period. For a rolling window, on the other hand, the length of the in-sample period is fixed. In this case the start date and end date successively increase by one observation.

Overall, reestimating the parameters on a daily basis improved the forecasting ability of the model. The results for a recursive and rolling window technique were similar, when the length of the rolling window was chosen to be at least nine months or longer. For shorter windows, the parameter estimates for the GARCH and in particular for the regime-switching models showed some instability. In the following only the results for the recursive window approach are provided. However, the results for the static and rolling window approach are available upon request to the authors.

For point forecasts we measure the average prediction errors by computing the mean absolute error (MAE) and mean squared error (MSE) of the one-day-ahead forecasts. The results for the different models can be found in Table 1.5. We observe the smallest MAE for the 'AR' model despite the superior in-sample fit of the GARCH or regime-switching model. On the other hand, the smallest MSE can be observed for the regime-switching model 'MR' with an autoregressive term in the base regime. We also find that for both criteria, the GARCH model yields the worst results. However, the differences between the results for all models are rather small, since the values for MAE range from 0.0306 to 0.0310 and for MSE from 0.0042 for the regime-switching model 'MR' to 0.0049 for the GARCH model. Overall, we conclude that for point forecasts the results are mixed while there are no substantial differences between the models.

Table 1.5: Point forecast results

The table reports results for the mean absolute error (MAE) and mean-squared error (MSE) for the point forecasts of the considered models.

	MAE	MSE
i.i.d. Normal	0.0307	0.0043
AR(1)	0.0306	0.0047
GARCH(1,1)	0.0310	0.0049
'Gaussian'	0.0308	0.0044
'MR'	0.0308	0.0042

In a second step we investigate the ability of the models to provide accurate forecasts of the whole density function or intervals. Especially for risk management purposes such forecasts are highly relevant, since traders and brokers are more interested in predicting intervals or densities for future price movements than in simple point estimates. The literature suggests different approaches to evaluate interval or density forecasts, see e.g. Christoffersen (1998); Christoffersen and Diebold (2000); Crnkovic and Drachman (1996); Diebold, Gunther and Tay (1998). One approach (Christoffersen, 1998) is to evaluate the quality of confidence interval forecasts by comparing the nominal coverage of the models to the true coverage in the out-of-sample period. However, tests being based on confidence intervals may be unstable in the sense that they are sensitive to the choice of the confidence

1.4. EMPIRICAL RESULTS

level α. We overcome these deficiencies by applying a test that investigates the complete distribution forecast, instead of a number of quantiles only. Evaluating the accuracy of the density forecasts we perform a distributional test following Crnkovic and Drachman (1996) and Diebold et al. (1998). We are interested in the distribution of the logreturn y_{t+1}, $t > 0$, which is forecasted at time t. Further, let $f(y_{t+1})$ be the probability density and $F(y_{t+1}) = \int_{-\infty}^{y_{t+1}} f(x)dx$ be the associated distribution function of y_{t+1}. To conduct the test, we determine $\hat{F}(y_{t+1})$ by using the parameter estimates from the in-sample period and the observations $y_s, s = 0, .., t$. Rosenblatt (1952) shows that if $\hat{F}$ is the correct loss distribution, the transformation of y_t, namely

$$u_{t+1} = \int_{-\infty}^{y_{t+1}} \hat{f}(x)dx = \hat{F}(y_{t+1}), \qquad (1.12)$$

is i.i.d. uniformly on $[0, 1]$. The method can be applied to test for violations of either independence or uniformity.

Figure 1.7 presents the corresponding probability integral transforms of the one-day ahead forecasts based on the 'naive' model of a simple normal distribution, the AR(1) model, the GARCH model and the regime-switching model specification 'MR'. It turns out that the observations for u_t of the models with a simple normal distribution and the AR(1) process for the logreturns are far from being uniformly distributed. A very high fraction of the probability integral transforms lies in the two central quartiles between 0.25 and 0.75, indicating that using a simple normal distribution or AR(1) model, very often the forecasted confidence intervals for the next day are too wide. This is also confirmed by Figure 1.8, displaying the observed logreturns and predicted 95%-confidence intervals for the different models from July 3, 2006 to December 29, 2006. For the GARCH and regime-switching models we obtain significantly better results. The corresponding probability integral transforms are closer to a uniform distribution. As Figure 1.8 indicates, the width of the confidence intervals varies with the conditional variance of the density forecast, such that during periods of higher volatility the intervals become wider. However, both for

Figure 1.7: One-day ahead forecast results

The figure displays histograms of the probability integral transforms of the one-day ahead forecasts for logreturns of CO_2 emission allowances for January 1 2006 - May 31, 2006. Results for the 'naive' model of a simple normal distribution (Upper left panel), AR(1) model (Upper right panel), the GARCH(1,1) model (Lower left panel) and the 'MR' regime-switching model (Lower right panel).

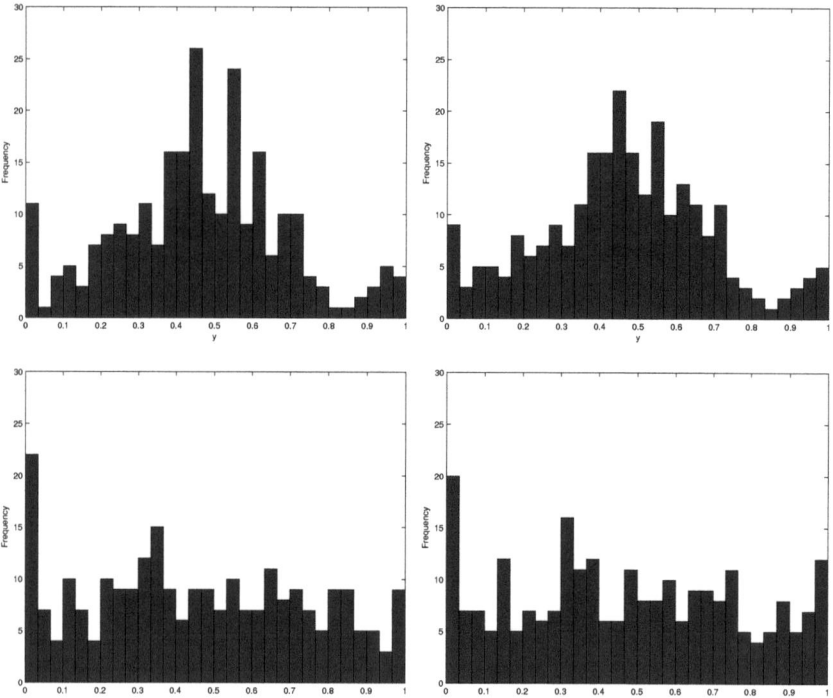

the GARCH and regime-switching models there is a high number of observations with probability integral transforms close to zero. This may be due to the fact that substantial price shocks like in April 2006 are rather difficult to predict with an econometric model. As a consequence, several of the large negative returns could not be captured despite the use of models with conditional variance and resulting wider confidence intervals for these

1.4. EMPIRICAL RESULTS

Figure 1.8: Logreturns of EUA prices and predicted 95%-confidence intervals

The figures display logreturns of EUA prices and predicted 95%-confidence intervals for the different models from July 3, 2006 to December 29, 2006. Results for the 'naive' model of a simple normal distribution (Upper left panel), AR(1) model (Upper right panel), the GARCH(1,1) model (Lower left panel) and the 'MR' regime-switching model (Lower right panel).

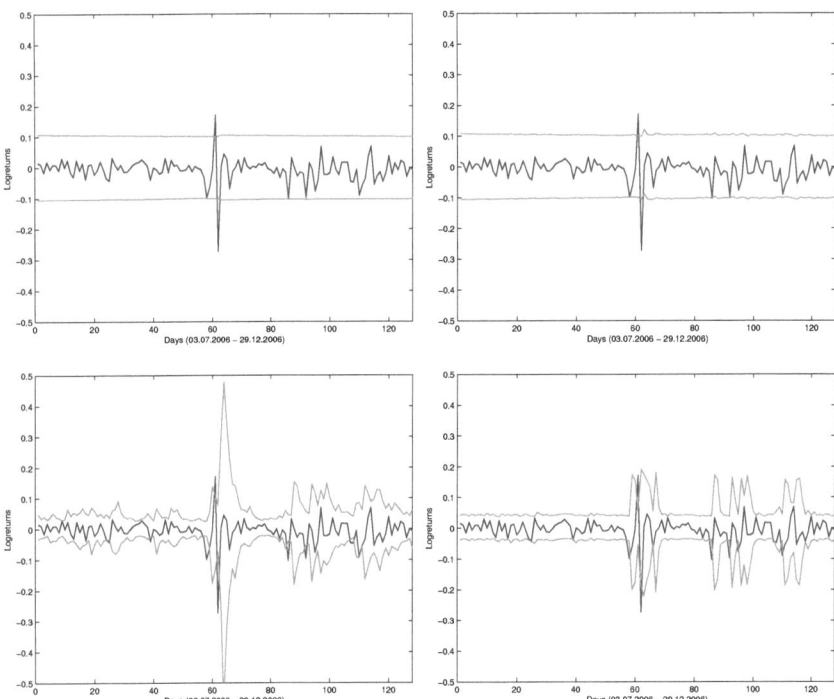

periods.

Testing for uniformity, Crnkovic and Drachman (1996) suggest to use a test that is based on the distance between the empirical and the theoretical cumulative distribution function of the uniform distribution. This may be done using e.g. the Kolmogorov-Smirnov (KS) or Kuiper statistic. The former is usually denoted by $D_{KS} = max\{D^+, D^-\}$ while the latter is

$D_{Kuiper} = D^+ + D^-$ with $D^+ = sup\{F_n(u) - \hat{F}(u)\}$ and $D^- = sup\{\hat{F}(u) - F_n(u)\}$. Hereby $F_n(u)$ denotes the empirical distribution function for the probability integral transforms of the one-day ahead forecasts and $\hat{F}(u)$ is the cdf of the uniform distribution.

Table 1.6 presents the test results for the models. We find, that the 'naive' model of a simple normal distribution for the logreturns gives the worst results. Probability integral transforms of the one-day ahead forecasts are non-uniformly distributed. Both tests reject the hypothesis of a uniform distribution even at the 1% level. Similar results are obtained for the AR model. The KS and Kuiper test statistics also significantly reject the assumption of uniformity at the 1% level. The results obtained for the GARCH and regime switching models are clearly superior. As indicated by Figure 1.7 the probability integral transforms are much closer to uniformity in comparison to the normal distribution and the AR model. For all three models, the assumption of uniformity cannot be rejected even at the 10% significance level. The best results for the KS test are obtained for the regime-switching 'MR' model. However, the results for the other regime-switching and GARCH model are only slightly worse. For the Kuiper test, the GARCH model outperforms all its competitors, but the distance for the two regime-switching models is in a similar range. So despite the fact that the GARCH and regime-switching models have some difficulties in forecasting a number of extreme negative price shocks, the density forecasts using these models seem to be adequate.

Overall, in terms of density forecasting, the GARCH and regime-switching models significantly outperform the models with constant variance. This suggests the models as particularly useful for risk management purposes and short-term forecasting of future price ranges for emission allowances.

1.4.6 Comparison with results from other papers

Daskalakis et al. (2006) suggests that CO_2 emission allowance price levels are non-stationary and exhibit abrupt discontinuous shifts. For logarithmic returns they find that the distribution is clearly non-normal and characterized by heavy tails. They further find that the

1.4. EMPIRICAL RESULTS

Table 1.6: Results for Kolmogorov-Smirnov and Kuiper statistics

The table reports the results for the Kolmogorov-Smirnov and Kuiper statistics. Best results are highlighted in bold. The asterix further denote rejection of the model at the 1% ***, 5% ** or 10% * level, for $n=253$ observations.

	KS	Kuiper
i.i.d. Normal	0.1760***	0.2620***
AR(1)	0.1750***	0.2591***
GARCH	0.0748	0.0828
'Gaussian'	0.0712	0.0893
'MR'	0.0709	0.0885

best model fit for allowance prices in terms of likelihood and parsimony is obtained by a geometric Brownian motion with an additional jump-diffusion component. This model is also able to produce the discontinuous shifts in the underlying diffusion that are observed in the CO_2 emission allowances prices. Although our approach differs from their analysis, we find a superior performance of the models with non-constant variance like GARCH or regime-switching confirming the non-normality and heavy-tails in the logreturns. The models not only provide the best in-sample fit but also outperform alternative approaches with constant variance in density and volatility forecasting. Hence, similar to Daskalakis et al. (2006), we find that issues like shifts in prices, non-normality or short periods of extreme volatility have to be incorporated into adequate pricing or forecasting models for CO_2 allowances or returns.

Paolella and Taschini (2006) examine the performance of different GARCH models for CO_2 and SO_2 certificates. Similar to our results, they observe heteroskedasticity in the returns and obtain an adequate fit for models with conditional variance. They conclude that for sound risk management, hedging or purchasing strategies the choice of an adequate statistical model is a crucial task. Finally, Seifert et al. (2008) develop a stochastic equilibrium model in order to analyze the dynamic behavior of CO_2 emission allowances spot prices for the European emissions market. According to their analysis, spot prices must always be positive and bounded by the penalty cost plus the cost of having to deliver any lacking allowances. As far as volatility is concerned, they argued that a steep increase will occur when the end of the trading period is approaching. This also recommends the use of models with conditional variance to capture the fact whether the market is in period of higher or lower volatility.

1.5 Summary and Conclusion

In this chapter we examine the spot price dynamics of CO_2 emission allowances in the EU ETS. Short-term dynamics of the new asset are of particular interest for market participants like risk managers or traders, but also for CO_2 emitting companies, as they must model the behavior of their production costs. We find that the logreturns exhibit skewness, excess kurtosis and different phases of volatility behavior coming from fluctuations in demand for CO_2 allowances. The best in-sample fit to the data is provided by a regime-switching model with an autoregressive process in the base regime and a normal distribution for the spike regime. Results for the GARCH and normal mixture regime-switching models are only slightly worse while the fit of the models with constant variance like a Gaussian distribution and an AR(1) process is clearly inferior. We also provide an out-of-sample forecasting analysis for the CO_2 allowance logreturns. In terms of point forecasts we observe only very small differences between the models for the evaluated MAE and MSE measures. We also conduct an analysis on density and interval forecasts which is more relevant for risk managers than simple point forecasts. For one-day ahead density forecasts, the AR-GARCH and regime-switching models clearly outperform the models with constant variance. The adequacy of a simple normal distribution or AR process is significantly rejected.

The superior performance of the models with conditional variance can be explained to a high extend by the relationship between allowance prices, regulatory factors and fundamental variables. In particular, political issues like the over-allocation of certificates or the following insecurity about whether the EU would decide to reduce the number of allowances granted to the industry should be named here: consequences were prices dropping

1.5. SUMMARY AND CONCLUSION

substantially and a following phase of high volatility what is captured much better by the GARCH and regime-switching models. Also periods of unexpected weather like cold snaps or extremely hot and dry summer months lead to phases of price behavior that favors the more flexible models with conditional volatility. The suggested models can be used in particular for Value-at-Risk purposes. Modeling the short-term price behavior of emission allowances will be especially helpful for risk managers, brokers or traders in the market, but might also enable companies to monitor the costs of CO_2 emissions in their production process. Our results strongly support the use of AR-GARCH or regime-switching models for modeling the returns of CO_2 emission allowances. The models may also be used for the pricing of relates derivatives on emission allowances. For further references on option and derivative pricing with GARCH and regime-switching models we refer e.g. to Huisman and De Jong (2003)and Duan (1995). Overall, we suggest further investigating the use of these models for this new class of assets in future work when more empirical data is available.

Chapter 2

Liquidity and Price Discovery in the CO_2 Futures Market: An Intraday Analysis

> European Union CO_2 allowances (EUAs) are traded on several markets with increasing intensity. We provide an intraday data analysis of the EUA futures market for the complete first trading period 2005-2007. To investigate the trading process in this young market, we compare the two main trading platforms ECX and Nord Pool with respect to liquidity and price discovery. Both are of high relevance to traders. We analyze liquidity by estimating traded bid-ask spreads following the approach of Madhavan, Richardson and Roomans (1997) and study price discovery using the VECM framework of Engle and Granger (1987). We find that while estimated transaction costs are always lower on the larger exchange ECX, the less liquid platform Nord Pool also contributes to price discovery, especially during the first months of trading. Overall, results indicate that from 2005 to 2007 liquidity in the European CO_2 futures market has markedly increased and organized trading has rapidly expanded.

2.1 Introduction

With the official start of the European Union Greenhouse Gas Emission Trading Scheme (EU ETS) in January 2005, a new European commodity market has been created. In the market for European Union Allowances (EUAs), purchasing one EUA entitles the holder to emit one ton of CO_2 equivalent greenhouse gases. With an increasing range of new instruments (e.g. spot, forwards, futures and options) the carbon market has steadily gained complexity. Currently, the EU ETS is the largest CO_2 trading scheme world wide.

While prices from the OTC market have served as reference prices at the beginning of the EU ETS, their importance has declined with the development of standardized carbon products on distinct trading platforms. During the first trading period (Phase 1), which lasted from 2005 to 2007, organized allowance trading has been fragmented across five trading platforms: European Climate Exchange (ECX), Nord Pool, Powernext, European Energy Exchange (EEX) and Energy Exchange Austria (EXAA). Since the underlying asset is equal on all exchanges, questions with respect to liquidity migration and price discovery across trading platforms are important factors to investigate.

The end of the first trading period of the EU ETS in December 2007 provides an excellent opportunity to address these questions and to give a comprehensive overview of the European carbon market development. Being the first to have access to intraday transactions data, we are able to complement the existing literature by investigating this very recent market from a microstructure angle. Since almost all trading takes place in the futures markets, we focus on futures price data supplied by ECX and Nord Pool, the two most liquid European trading exchanges for futures EUAs. The aim of this chapter is twofold. Apart from providing an overview of the development of trading on both exchanges, we compare two microstructure issues that are of high relevance to potential traders: liquidity and price discovery. With respect to liquidity, we start by comparing overall trading volumes as well as the development of trading frequencies across exchanges. We then estimate traded bid-ask spreads following the approach of Madhavan et al. (1997) that allows the estimation of spreads when no quote data but only transaction data and a trade indicator variable are available. Applying this procedure has the advantage that it enables us to infer main causes of trading frictions and, hence, transaction costs. To analyze relative price discovery on both exchanges we use the VECM framework by Engle and Granger (1987) building on the cointegration relationship between transaction price series. To quantify the two markets' relative contributions to the price discovery process we apply two different measures: common factor weights proposed by Schwarz and Szakmary (1994) and information shares as introduced by Hasbrouck (1995).

Our analysis is related to a vast body of market microstructure literature that investigates

2.1. INTRODUCTION

liquidity and price discovery on financial markets. Regarding the European carbon market, there is no literature analyzing bid-ask spreads. A few studies have addressed the question of price discovery between spot and futures markets (Seifert, Uhrig-Homburg and Wagner, 2006; Daskalakis et al., 2006; Milunovich and Joyeux, 2007), but no work has investigated price discovery between futures prices on distinct exchanges. Since almost all trading volume takes place in futures markets, we believe our study to be of high relevance. Furthermore, the mentioned studies only use daily data, possibly blurring results of price leadership if price discovery takes place at finer trading intervals.

We believe that our results are of interest for regulatory authorities that are in charge of the design of the upcoming commitment periods, for operators of exchange platforms, for researchers interested in the application of microstructure tools to new markets and, equally important, for agents who trade actively in the market like market makers, brokers, arbitrageurs, etc. It is also possible to use our results in order to evaluate the relative development of the markets. From the public, a lot of criticism has been raised about the market in Phase 1 mainly due to the significant over-allocation of EUAs. Academic work like Daskalakis and Markellos (2007) and Milunovich and Joyeux (2007), using data until the end of 2006, conclude that weak form informational efficiency in the European CO_2 market is violated. However, Uhrig-Homburg and Wagner (2007) found evidence in favor of a cost-of-carry pricing mechanism for futures expiring within Phase 1 of the market. Our evidence shows that trading frictions in forms of transaction costs have decreased over the first trading phase, trading volume has increased and price discovery takes place across exchanges. Hence, it appears that from a trading perspective, the market has made a lot of progress since its operational start in January 2005.

The remainder of the chapter is organized as follows. Section 2.3 introduces the reader briefly to the organization of the European carbon market and to the institutional details that are relevant for the data collection procedure. Section 2.4 describes the methodology of the bid-ask spread analysis and its econometric application. The price discovery process using an error correction model is explained in Section 2.5. Estimation results for both types of analysis are displayed subsequent to the description of the methodology. The chapter ends with an interpretation of the results and a conclusion with respect to the future of the organized carbon market in Section 2.6.

2.2 Market Structure and Data

2.2.1 Market Structure of Carbon Exchanges

2.3 Market Structure and Data

2.3.1 Institutional Background

The EU ETS started in January 2005 as a central instrument for member states of the European Union to achieve the emission reduction targets of the Kyoto Protocol in a cost-effective way.[1] It covers over 10,000 installations in the energy and industrial sectors that are collectively responsible for about 50% of European CO_2 emissions. Trading is organized in several stages. The first trading period served as a pilot phase and covers the years 2005-2007 while the second trading period from 2008-2012 constitutes the Kyoto commitment period (Phase 2). Plans for the post Kyoto trading period 2013-2020 (Phase 3) became more concrete after the United Nations summit in Bali in December 2007. Besides, in January 2008 the European Commission has agreed on a so called "Climate and Energy Package", which makes first regulatory suggestions and improvements for the continuation of action against climate change in the EU.

The EU ETS is organized as a cap-and-trade scheme where participating firms have to reduce the amount of emitted CO_2 and annually demonstrate that their level of EUAs corresponds to their actual emissions. Every year, at the end of February, a certain amount of EUAs is allocated to the compliant firms for the current trading year according to National Allocation Plans (NAPs). On April 30 of the following year, firms have to deliver the required EUAs to the national surveillance authorities according to their actual emissions

---

[1] On an EU-wide level, emissions have to be reduced by 8% in the first Kyoto commitment period 2008-2012 relative to the output level of 1990.

2.3. MARKET STRUCTURE AND DATA

volume. Not handing in the required amount of emissions is fined with an extra fee of Euro 40 (Euro 100) per missing EUA in the pilot period (Phase 2) additional to delivering the missing amount of EUAs.

Companies being able to keep emissions below their allocation level are free to sell excess allowances in the market. Firms which need additional allowances to comply with their output levels have the choice to either invest in emissions-reducing technologies, to switch to less emissions-intensive production technologies or, if marginal abatement costs are higher than the market price of EUAs, to buy EUAs on the European CO_2 market.

Within Phase 1 and Phase 2 surplus allowances can be transferred for use during the following year (banking). Banking between Phase 1 and Phase 2 was forbidden by most of the countries. Only France and Poland allowed for restricted banking. As allocation always takes place in February, borrowing of EUAs from the future year is indirectly possible as the compliance date for the preceding year is April 30. However, it was not possible to borrow EUAs between 2007 and 2008.[2] Trading is organized as bilateral, over-the-counter (OTC) and organized exchange trading. It takes the form of agency or proprietary trading and may be for compliance, speculative or arbitrage purposes.

To get an overview of how many allowances have been exchanged among market agents, Table 2.1 displays the total trading volumes split into futures and spot activities since the EU ETS has been operating and includes both OTC and exchange trading. It can be seen that overall trading volume markedly increased from 121 Mio tons of CO_2 in 2005 to 1 123 Mio tons of CO_2 in 2007. The share of spot relative to overall trading volume declined from 8.5% to 5.4%.

2.3.2 Market Structure of Carbon Exchanges

In Phase 1, organized EUA trading took place at five exchange platforms. ECX only offers futures, Powernext and EXAA only offer spot trading whereas on EEX and Nord Pool both instrument types can be traded. In the following analysis we focus on the

---

[2]Note that consequently there exist essentially two spot markets, one for Phase 1 and one for Phase 2 (Seifert et al., 2006).

Table 2.1: Overall trading volume of the EUA spot and futures market in Phase 1

The table depicts overall trading volume of the EUA spot and futures market in the Phase 1 (2005-2007) in Million tons of CO_2.

Year	Spot [Mio t of CO_2]	Futures [Mio t of CO_2]
2005	10.25	110.82
2006	49.53	508.29
2007	60.26	1 062.42

Source: Own calculations.

two main trading venues ECX and Nord Pool which comprise by far the largest exchange traded futures volume: In 2006, ECX being a member of the Climate Exchange Plc group possessed a market share of 86.5%. The Norwegian platform Nord Pool had a share of 12.5%, see also Daskalakis et al. (2006). In terms of overall market share, in early 2007 ECX accounted for 56% of EUA trading volume, being followed by OTC trading volume with 42%.

The traded futures instruments on both platforms are standardized contracts giving the holder the right and the obligation to buy or sell a certain amount of EUAs at a certain date in the future at a pre-determined price. On both exchanges, one futures contract ('lot') corresponds to 1 000 EUAs and hence delivers the right to emit 1 000 tons of CO_2 equivalent. The contracts allow to lock in prices for delivery of EUAs at given dates in the future with delivery guaranteed by the respective clearing house. Counterparty risk is mitigated by specific margin requirements. The contracts are supposed to facilitate trading, risk management, hedging and physical delivery of EUAs. While contracts with monthly expiry and annual contracts with expiry in March exist, we focus on the by far most liquid annual contracts with expiry in December. These contracts expire on the first business day of December on Nord Pool and on the last Monday of December on ECX.[3] Settlement is three days after the last trading day.

---

[3]If there is a public holiday in the respective trading week, the prior Monday is taken. The procedure continues until there is no public holiday in the trading week.

2.3. MARKET STRUCTURE AND DATA

On both exchanges, trading is organized as continuous trading and takes place anonymously on electronic platforms. Exchange hours are from 08:00 to 18:00 CET on ECX and from 08:00 to 15:30 CET on Nord Pool.[4] On ECX the trading period is preceded by a pre-opening session from 07:45 to 07:59 CET. No actual trading takes place during this period, traders can only input orders that they wish to execute once trading begins at 08:00 CET. The daily closing period lasts from 17:00 to 17:15 CET. On Nord Pool, daily closing prices are determined between 15:20 and 15:30 CET at a randomly selected point in time. On both exchanges trading is not interrupted by intraday auctions. Currently, ECX has 92 and Nord Pool has 97 members engaging in EUA trading.

With respect to order processing, both exchanges do not show markable differences. Incoming orders are binding until the end of the trading day if they have not been executed, changed or canceled. Order types include order book (limit) orders, market orders and stop orders. Matching occurs according to price and time priority. Initially, the minimum tick size for ECX futures was Euro 0.05 per CO_2 emission allowances. Since March 27, 2007 it is reduced to Euro 0.01. On Nord Pool the minimum tick size is always Euro 0.01 per CO_2 emission allowance. Currently, trading and clearing fees per contract amount to Euro 3.50 on ECX and to Euro 3.00 on Nord Pool for members.[5] The annual fee for full members is Euro 2 500 on ECX and Euro 3 000 on Nord Pool.

Both exchanges have introduced market makers to boost liquidity. Until June 18, 2006, EdF Trading Limited was active as a market maker for the EUA market on Nord Pool. On January 9, 2007, the new market maker Alfa Kraft AB started to operate. As a minimum requirement, market makers have to quote prices from 08:30 to 10:00 CET and from 13:00 to 15:30 CET. Before 2007, the minimum quoting periods were from 10:30 to 12:00 and from 14:00 to 15:30. Restrictions with respect to maximum spreads and minimum volumes apply. While the minimum offered volume is 5 000 tons, maximum spreads are EUR 0.50

---

[4] On Nord Pool, trading hours were extended in June 2005 from 09:00 (10:00) to 15:30 CET in March (February) 2005.

[5] On ECX, trading and clearing fees, as for e.g. order routing customers and client business, amount to Euro 4.00 per contract. Note that fees have decreased over time. For instance on Nord Pool trading and clearing fees amounted to Euro 70.00 per contract at the launch of EUA trading before gradually declining to Euro 3.00 in December 2006.

for the nearest December contract and EUR 0.75 for the following December contract. As another method to enhance liquidity and to promote electronic trading, Nord Pool has launched a so-called initiator/aggressor fee model in January 2006. Electronically incoming quotes from "initiators" can be executed free of trading charges such that only clearing fees remain. "Aggressors" (price takers) have to pay the ordinary trading fees.

ECX launched a market maker program in July 2007. Currently, Fortis Bank Global Clearing and Jane Street Global Trading are active as market makers. Requirements are stricter compared to Nord Pool in that both bid and ask prices must be quoted for at least 85% of the trading time between 09:00 to 18:00 CET. Spreads may not exceed Euro 0.15 for the December 2008 contract and Euro 0.25 for the other Phase 2 contracts. The minimum quoting volume corresponds to 10 000 tons. Market makers have to respond to quote requests within five minutes.[6]

Overall, we believe that differences in the market organization of both markets do not allow us to predict a clear pattern for differences in bid-ask spreads and price discovery across exchanges. The main features of the trading process are similar on both markets. Regarding differences, traders on Nord Pool have to pay lower transaction costs compared to ECX which may however be offset by a higher annual fee. Furthermore, while we could not obtain historical fee data for ECX, we know that trading costs on Nord Pool were markedly higher in the beginning. Hence, it may be the case that trading fees on ECX were lower compared to Nord Pool before the end of 2006. While market makers have been introduced earlier on Nord Pool than on ECX, maximum applicable spreads are markedly tighter on ECX. In the initiator/aggressor model trading costs are waived for the initial price quoters. This may attract additional liquidity suppliers on Nord Pool whose competition might narrow spreads on this platform. Finally, the possibility to trade longer on ECX may be one reason to favor trading on ECX over trading on Nord Pool. The data reveal that 20% of daily ECX trades occur within the last 2.5 trading hours (from 15:30 to 18:00 CET), that is after trading on Nord Pool has finished.

---

[6]Information is obtained from the official websites www.europeanclimateexchange.com and www.nordpool.no in July 2008.

2.3. MARKET STRUCTURE AND DATA

2.3.3 Data Set and Summary Statistics

In the following analysis we use intraday transaction data for annual standardized EUA futures and forward contracts being traded from April 22, 2005 to December 28, 2007 on ECX and from February 11, 2005 to December 28, 2007 on Nord Pool. After providing some summary statistics to show the market development on both platforms, we briefly address some data collection issues that result from investigating data from two distinct markets.

Table 2.2 depicts trading volumes (without OTC) of EUA futures with expiry in December for both platforms disaggregated by years and contracts. It can be seen that trading volume has markedly increased both over years and over contracts, but to a higher extent on ECX compared to Nord Pool. Highest trading activity takes place in the nearby futures contract. The development of the Dec07 contract in 2007 is an exception and is due to the publication of the large EUA over-allocation in April 2006, which led to a marked price decline (compare also Figure 2.1 showing the development of futures prices on ECX for Phase 1). Due to lack of liquidity, for the rest of the chapter we only consider the Dec05 to Dec08 contracts and disregard those with later expiry, i.e. the Dec09 to Dec12 contracts.

Figures 2.2 and 2.3 depict the daily transaction frequencies and the average monthly standard deviation of daily returns for each platform and contract. Again, it can be inferred that transaction frequencies are highest for the nearby contract and hence markedly increase in December of the year prior to expiry. If we compare volatility across contracts and exchanges we observe high volatility in the market where the respective contract is launched first. For the Dec05 and Dec06 contracts there are two peaks that disturb the relatively smooth volatility pattern; one in July 2005 for the Dec05 contract and one in April/May 2006 for Dec06 contract. The first one probably relates to large price fluctuations as a consequence of unexpected selling in the market by some Eastern European countries that succeeded to obtain access to the carbon market earlier than anticipated. Maybe it also reflects overall uncertainty with respect to price drivers as a consequence of the terrorist attacks in London in July 2005. The second peak can be clearly linked to the

Table 2.2: trading volumes (without OTC) of EUA futures with expiry in December

The table depicts trading volumes (without OTC) of EUA futures with expiry in December on ECX and Nord Pool broken down into contract and year.

Year	Contract	ECX		Nord Pool	
		Mio t of CO_2	Mio Euro	Mio t of CO_2	Mio Euro
2005	Dec05	22.96	522.79	6.76	139.89
	Dec06	6.78	151.88	1.91	41.86
	Dec07	1.93	43.91	1.05	23.39
	Dec08	0.49	10.46	0	0
	Dec09-Dec12	0.02	0.43	0	0
	Sum	32.18	729.47	9.72	205.14
2006	Dec06	93.77	1 763.93	9.92	182.47
	Dec07	35.25	520.51	1.25	19.66
	Dec08	29.56	540.10	0.37	6.87
	Dec09-Dec12	0.46	9.16	0	0
	Sum	159.04	2 833.7	11.54	209
2007	Dec07	50.24	65.22	3.10	4.03
	Dec08	258.96	5 237.18	18.38	380.78
	Dec09-Dec12	29.44	652.86	0.25	5.41
	Sum	338.64	5 955.26	21.73	390.22

market breakdown when the significant over-allocation of EUAs became public at the end of April 2006.[7] Consequently, the Dec07 contract became worthless to the firms as they were not allowed to transfer excess EUAs from 2007 into 2008. This price decline translates into high return volatility in the year 2007, especially in the last months of trading when price variations of 0.01 Euro were very high compared to a price level of about 0.03 Euro. Finally, for the Dec08 contract, except for the high volatility at the launch of the contract at both platforms, the volatility pattern is smooth. Comparing the two figures, it can be observed that often trading intensity and market volatility move in the same direction.

---

[7]Compare the weekly newsletter at www.climatecorp.com.

2.3. MARKET STRUCTURE AND DATA

Figure 2.1: EUA futures prices in Phase 1

The figure displays EUA futures prices for the Dec05 to Dec08 contracts in Phase 1 on ECX.

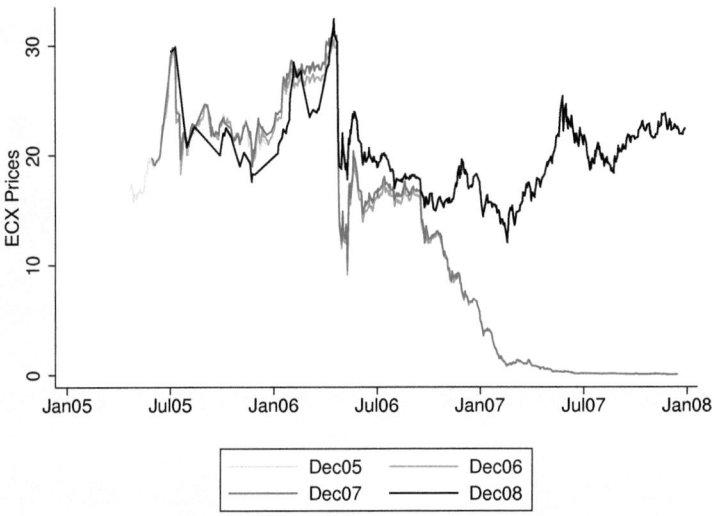

Data Collection

When investigating the development of bid-ask spreads and price discovery on both exchanges, we have to address some issues related to the differences in trading protocols and contract specifications as well as some standard high frequency issues. To start with, for both types of analysis we omit overnight returns that could induce heteroskedasticity into our data set. Then, for comparing bid-ask spreads across exchanges over the whole trading period of contracts, we omit non-overlapping trading intervals and focus only on periods when the respective contract was traded on both exchanges. We furthermore aggregate all trades within the same second that have the same trade indicator to account for price effects of orders walking up or down the book. Finally, we only include data from continuous

Figure 2.2: Monthly transaction frequencies in Phase 1

The figure displays monthly transaction frequencies for the Dec05 to Dec08 contracts in Phase 1 on Nord Pool (top panel) and ECX (bottom panel).

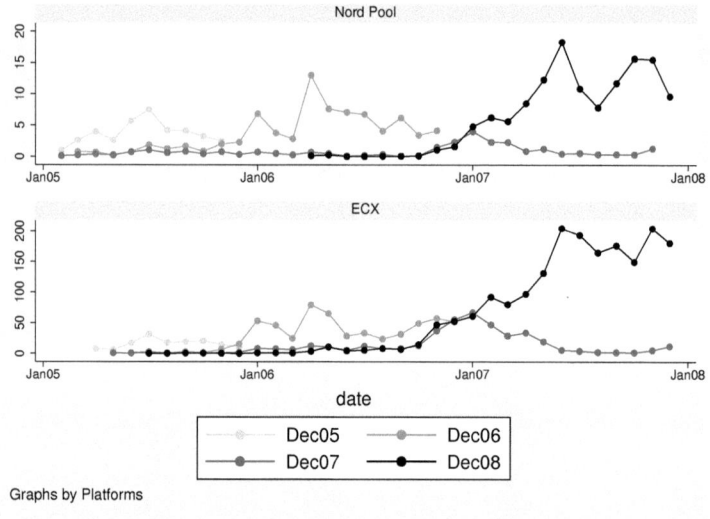

trading periods into our analysis and hence exclude pre-opening and post-closing prices. Regarding the price discovery analysis, some further re-organization of the data set is necessary prior to estimation. First, in order to synchronize trading hours we delete ECX trades that occur after 15:30. Thus, throughout the chapter we only use data from 08:00 until 15:30 CET. Second, in our price discovery analysis we postulate a one-to-one relationship between the prices of futures contracts traded on different markets. However, that relationship does not exactly hold due to differing expiration dates on both exchanges. Hence, we discount all contracts to their present value at the respective trading day.[8] The third

---

[8] We mainly use monthly interest rates for discounting and linearly interpolate interest rates between months. Interest rate data is obtained from Datastream. For very short discounting horizons, we use EONIA interest rates, for horizons up to a year we use Euribor interest rates and for horizons of more

2.3. MARKET STRUCTURE AND DATA

Figure 2.3: Monthly average return standard deviation

The figure displays monthly average return standard deviation at Nord Pool and ECX for the respective trading period of each contract, Dec05 to Dec08.

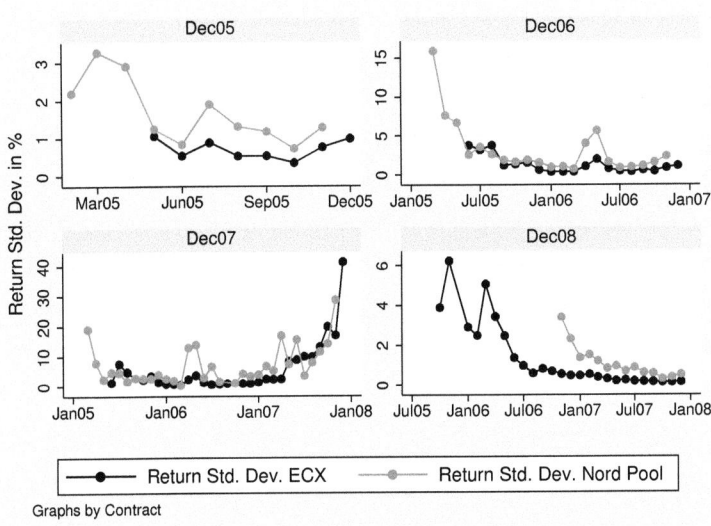

and probably most important aspect that has to be considered is the question of price synchronization. Transactions do not occur at regular intervals, nor do transactions in the two parallel trading platforms take place simultaneously. Explicitly, our data includes much more transaction prices for ECX than for Nord Pool. Thus several futures prices have to be eliminated from the series. To synchronize the two price series we form three different data sets of matched trade pairs. First, beginning at the start of each trading day, for every transaction price on Nord Pool we identify the most recent transaction price at ECX. These pairs are saved and generate the price series for our model (NP-Match). This method favors Nord Pool, the less liquid exchange. Estimation results that systematically

---

than one year we employ European monthly corporate interest swap rates.

use "stale" prices of ECX are likely to underestimate the role of ECX and are hence a very conservative measure for price leadership of ECX. If, contrary to expectations, we find a large share of price discovery on Nord Pool, these results might stem from the fact that we systematically favor Nord Pool. It is easy to inquire the robustness of these findings by applying an analogous synchronization procedure favoring ECX (ECX-Match). A third possibility not clearly favoring one exchange over the other has been suggested by Harris, McInish, Shoesmith and Wood (1995).[9] The authors synchronize the data as follows. Beginning at the start of each trading day, as soon as a trade has taken place on both exchanges the trade which has occurred latest in time is matched with the most recent trade on the other exchange. This pair is saved and a new matched trade pair is formed in the same manner for the whole data sample (Harris-Match).[10] Obviously, for the Harris-Match the frequency of the data is determined by the market with the fewest trades, which is in our analysis Nord Pool.

Since we expect price discovery to take place at the larger and more liquid trading platform ECX, we opt for choosing the NP-Match disfavoring ECX as a benchmark and refer to other matching algorithms as a robustness check.

2.4 Spread Analysis

In this section, we investigate the development of transaction costs on both exchanges for Phase 1. We measure transaction costs by estimating bid-ask spreads, which are defined as the difference between the best quoted ask and the best quoted bid price in the market. This measure can be interpreted as the costs of trading a round-trip (i.e. costs paid by a liquidity demander to a liquidity supplier for an instantaneous buy and sell transaction) or

---

[9] Another possibility of synchronization is the use of equidistant time intervals. This procedure consists of matching the last observed prices at the end of pre-specified time intervals (e.g. 5 minutes, 30 minutes) in each market. If no price updating has taken place within one time interval, the most recent price in the respective market is used for the matching. In our analysis we do not consider this matching approach as the probability that the most recent price comes from the more liquid market ECX is high. Consequently, ECX would be favored compared to Nord Pool as a relatively new ECX trade would be probably matched with an older Nord Pool price.

[10] Compare Harris et al. (1995), pp. 566. This matching procedure is referred to as "REPLACE ALL".

2.4. SPREAD ANALYSIS

alternatively as the price concession that it takes to induce an agent waiting in the market to transact immediately instead of waiting until prices move in her favor. The existence of a bid-ask spread and hence of trading frictions is typically explained with the existence of order processing costs, inventory costs or asymmetric information costs.[11]

As examples for the latter in the carbon market, one can think of firms' private decisions that concern for instance the start-up, closure or expansion of new and old installations as well as private news about market entrants. Furthermore, market participants might have different incentives and possibilities to acquire information about e.g. market developments (current and future market scarcity, abatement costs and potential of other firms) and about regulatory issues (National Allocation Plans for upcoming trading phases, development of CER market, incorporation of other trading schemes into the EU ETS). It is reasonable to assume that big companies that are more affected by the ETS have better sources of information than smaller firms. This asymmetric information and the related uncertainty influence market scarcity and thus the market price for EUAs or project based EUAs (CERs).[12]

Since we do not have access to best ask and bid quotes, bid-ask spreads cannot be calculated immediately from the data. Fortunately however, the market microstructure literature has proposed a variety of procedures to estimate the spread and its components. Given that our data set identifies transactions as either buyer-initiated or seller-initiated, we can use a so called trade indicator model to estimate and compare traded spreads on ECX and Nord Pool.

---

[11]Order processing costs include costs like telecommunications costs or exchange fees that have to be paid by the liquidity provider. Inventory costs arise for risk-averse liquidity suppliers that bear the risk of having to build up unwanted inventory positions to accommodate public order flow. Asymmetric information costs arise if traders with private information are active in the market and trade on their information. In order to balance losses to informed traders, liquidity providers may charge a spread. Compare also the surveys by Madhavan (2000) and Biais, Glosten and Spatt (2005).

[12]Additional to EUAs, a market for Certified Emissions Reductions (CERs) - assets, which arise from energy-reducing projects in developing countries - has been created.

2.4.1 Methodology

Trade indicator models assume that information about the underlying asset is contained in the order flow. They use this variable in form of a binary trade initiation indicator to model short-run dynamics of quotes and transaction prices and to estimate traded spreads. Trade indicator models have been proposed by e.g. Glosten and Harris (1988), Huang and Stoll (1997) and Madhavan et al. (1997). Since we observe a significant degree of autocorrelation in our trade initiation variable, we opt for the GMM-approach suggested by Madhavan et al. (1997) that does not restrict autocorrelation in the order flow to be zero. One potential drawback of this approach is the assumption of a constant trade size. Since median trade size is equal across exchanges, we believe that the model can be applied to our setup.

Let P_t denote the transaction price of our underlying futures contract at time t, x_t is a trade indicator variable with $x_t = 1$ if the transaction at time t is buyer-initiated and $x_t = -1$ if it is seller-initiated.[13] We assume that purchases and sales are (unconditionally) equally likely, so that $E[x_t]=0$ and $\text{Var}[x_t]=1$. We assume that beliefs about the asset value might change due to new public information announcements that are not associated with the trading process and due to the order flow that provides a noisy signal about the future value of the underlying asset. The innovation in beliefs between $t-1$ and t from dissemination of public information is denoted by η_t which is an i.i.d. random variable with mean zero and variance σ_η^2. Buy (sell) orders are considered as a noisy signal about an upward (downward) revision in beliefs given that there are some traders with private information in the market. We assume that the revision in beliefs (or price impact) $\theta \geq 0$ is positively correlated with the innovation in order flow $x_t - E[x_t|x_{t-1}]$, such that the change in beliefs due to order flow is $\theta(x_t - E[x_t|x_{t-1}])$. Finally, let μ_t stand for the post-trade expectation of the "true" value of the stock conditional on public information and on the

---
[13]Compare Madhavan et al. (1997), pp. 1039.

2.4. SPREAD ANALYSIS

information revealed by the trade initiation variable. μ_t evolves according to

$$\mu_t = \mu_{t-1} + \theta(x_t - E[x_t|x_{t-1}]) + \eta_t. \tag{2.1}$$

We assume that the price generating process P_t is determined from the unobserved process (2.1) by adjusting for the costs of providing liquidity services ϕ_t (order processing costs, baseline inventory costs or mark-ups from non-competitive pricing). ϕ captures the non-permanent (transitory) effect of order flow on prices. Quotes are ex-post rational and are conditional on the trade initiation variable being a buy or a sell order, such that a bid-ask spread emerges ($P_t^a = [P_t|x_t = 1] > P_t^b = [P_t|x_t = -1]$):

$$P_t = \mu_t + \phi x_t + \xi_t = \mu_{t-1} + \theta(x_t - E[x_t|x_{t-1}]) + \phi x_t + \eta_t + \xi_t, \tag{2.2}$$

with ξ_t being an independent and identically distributed random variable with mean zero. To estimate Equation (2.2), we need to make assumptions about the dynamic behavior of the order flow. We assume a general Markov process for the trade indicator variable where $\gamma = Pr[x_t = x_{t-1}|x_{t-1}]$ denotes the probability that a trade at the ask (bid) follows a trade at the ask (bid). Positive serial correlation in the order flow arises for a variety of reasons such as the breaking up of orders or price continuity rules, leading to $\gamma > 0.5$. Let ρ denote first order autocorrelation of the stationary trade indicator variable x_t, i.e. $\rho = E[x_t x_{t-1}]/Var[x_t]$. It is straightforward to show that $\rho = 2\gamma - 1$ such that autocorrelation in the order flow is an increasing function of the probability of a continuation. In order to estimate Equation (2.2), we need to compute $E[x_t|x_{t-1}]$, i.e. the conditional expectation of the trade initiation variable given public information. It can be easily seen that $E[x_t|x_{t-1}] = \rho x_{t-1}$.[14] Now we only have to substitute out the unobservable belief μ_{t-1} of Equation (2.2) to obtain an estimable equation. We can do so by noting that

---

[14]Since $E[x_t|x_{t-1} = 1] = Pr[x_t = 1|x_{t_1} = 1] - Pr[x_t = -1|x_{t_1} = 1] = \gamma - (1-\gamma) = \rho$ and analogously $E[x_t|x_{t-1} = -1] = -\rho$.

$\mu_{t-1} = P_{t-1} - \phi x_{t-1} - \xi_{t-1}$ and obtain

$$P_t - P_{t-1} = (\phi + \theta)x_t - (\phi + \rho\theta)x_{t-1} + e_t, \qquad (2.3)$$

where $e_t = \eta_t + \xi_t - \xi_{t-1}$. In the absence of asymmetric information and transaction costs, the price follows a random walk process. In the presence of frictions, movements in the price P_t reflect order flow and noise induced by price discreteness as well as public information news. From Equation (2.3), we see that the implied bid-ask spread at time t is equal to $P_t^a - P_t^b = 2(\phi + \theta)$.[15]

2.4.2 Estimation Approach

Analog to Madhavan et al. (1997), we estimate Equation (2.3) by GMM as an elegant way to account for autocorrelation of the error term and for possible conditional heteroskedasticity. We estimate the model using standard orthogonality conditions and make use of the definition of the autocorrelation parameter $\rho = E[x_t x_{t-1}]/Var[x_t]$ as an additional constraint to separately identify our two parameters of interest θ (asymmetric information component) and ϕ (transitory spread component).[16]

2.4.3 Estimation Results

This section contains the results of GMM estimations of Equation (2.3) for the different contracts traded on ECX and Nord Pool. In order to obtain comparable results for the complete trading periods, we estimate the model using observations starting from the calender month, in which we have observations for both exchanges until the last common trading day of the contracts. As stated before, we also exclude overnight returns.[17] We furthermore only report results for estimations with at least 100 observations. To get a first intuition on liquidity at both exchanges at the contract level, Table 2.3 provides an

---
[15] $P_t^a - P_t^b = (\phi + \theta) \cdot 1 - (\phi + \theta) \cdot (-1) = 2(\phi + \theta)$.
[16] Note that the GMM estimation parameters are identical with OLS parameters.
[17] The exclusion of overnight returns drastically reduces the number of observations especially for the least liquid Dec07 contract on Nord Pool. Results from including overnight returns are similar.

2.4. SPREAD ANALYSIS

Table 2.3: Estimated half spreads for the common sample periods

The table depicts estimated half spreads $\hat{s}/2 = \hat{\phi} + \hat{\theta}$ in Euro and percent for the four contracts on ECX and Nord Pool. Results are obtained by GMM estimation of Equation (2.3) under the given moment conditions. Estimation periods for the Dec05, Dec06, Dec07, Dec08 contracts are 05/01/2005-12/01/2005, 07/01/2005-12/01/2006, 06/01/2005-12/03/2007 and 05/01/2006-12/28/2007, respectively.

Contract	ECX			Nord Pool			t-stat.
	$\hat{s}/2$	Adj. R^2	Obs.	$\hat{s}/2$	Adj. R^2	Obs.	
Dec05	0.0624	0.21	2 256	0.0750	0.18	501	-0.98
Dec06	0.0531	0.17	8 011	0.0877	0.09	1 266	-3.06***
Dec07	0.0323	0.06	5 197	0.0487	0.12	296	-1.79*
Dec08	0.0284	0.17	23 482	0.0582	0.25	2 248	-9.38***

overview of estimated half spreads in Euro, $\hat{s}/2 = \hat{\phi} + \hat{\theta}$, for each instrument estimated over the whole common sample period.

It can be seen that for each contract estimated half spreads on ECX are significantly lower than on Nord Pool except for the Dec05 contract.[18] As 2005 was the initial trading year for both trading platforms, the finding of no significant differences in the first traded contract, Dec05, might not be surprising. The subsequent gradual decrease in spread magnitude for the following contracts on ECX is consistent with a maturing and expanding market. Interestingly, the pattern is slightly different on Nord Pool. While for the Dec05 contract spreads are of similar magnitudes, the relative distance increases over Phase 1. Additionally, absolute estimated half spreads do not monotonically decrease over the differing contracts. Since it seems plausible to detect more frequent price updating on the exchange with lower bid-ask spreads, we expect ECX to be the leader with respect to price discov-

---

[18] As estimates are based on non synchronized data, i.e. on different trading frequencies, it is not possible to directly compare spread magnitudes in a statistical sense. In order to be able to make a statement, we conduct a t-test of equality of estimated half spreads across exchanges. For the t-test, we assume independent samples with a different sample size and variance. The t-statistic is computed as

$$t = \frac{\hat{s}/2^{ECX} - \hat{s}/2^{NordPool}}{\sqrt{\hat{\sigma}^2_{ECX} + \hat{\sigma}^2_{NordPool}}},$$

where $\hat{\sigma}^2$ is the variance of the estimated half spread for the respective market. *,**, and *** denote statistical rejection at the 10, 5, and 1 percent levels, respectively.

ery, the second part of our study. However, if differences in bid-ask spreads stem from the presence of a higher probability of informed trading on Nord Pool relative to ECX, results might be the other way around.[19]

To improve our understanding about how liquidity measured by traded bid-ask spreads has developed over time we subdivide the estimation periods into finer time intervals. Table 2.4 shows the development of estimated half spreads in Euro and percent for the most liquid trading year and its calendar quarters. In case that there are less than 100 observations, the preceding month is included into the analysis as indicated at the bottom of the table. If there are still not enough observations, no estimation results are reported.

At the very beginning of trading (Q2 and Q3 of 2005), the relative difference of spreads for the Dec05 contract on Nord Pool and ECX is small. In the fourth quarter (Q4) of 2005 spreads start to differ significantly. Estimated transaction costs on ECX are lower than on Nord Pool, except for the Dec07, the least liquid futures contract. It can be observed that spreads vary over time on both exchanges. The decreasing trend in absolute terms for all contracts is only broken twice. A temporary increase in the third quarter 2005 might be linked to the surge in oil and gas prices related to damages caused by the hurricanes Katrina and Rita in September 2005 and it may be related to the volatility increase in the market in July 2005. The increase in the second quarter (Q2) 2006, the quarter with the highest absolute and percentage spreads for all contracts traded at that time, can be clearly linked to the market breakdown when the significant over-allocation of EUAs became public at the end of April 2006.[20] Percentage half spreads that are reported in columns 4 and 8 of Table 2.4 generally move in line with absolute spread magnitudes. Only for the Dec07 contract percentage spreads are increasing while absolute spreads decrease since the market price fell to almost zero.

As our model allows us to decompose the estimated spreads into an asymmetric information (θ_t) and a transitory component (ϕ_t), Table A.12 in Appendix A.1 provides the

---

[19]Compare also Hasbrouck (1995), p. 1184.

[20]Quarterly results for ECX for the Dec07 and Dec08 contract are available from the authors upon request.

2.4. SPREAD ANALYSIS

Table 2.4: Estimated half spreads for the most liquid years and its quarters

The table depicts estimated half spreads $\hat{s}/2 = \hat{\phi} + \hat{\theta}$ in Euro and percent for the four contracts on ECX and Nord Pool by the most liquid years and by its calendar quarters. Results are obtained by GMM estimation of Equation (2.3) under the given moment conditions. Percentage spreads are obtained by dividing the estimated half spread by the median price level of the estimation period. Estimation periods are as indicated with e.g. Q3 2006 denoting July to September 2006. For Nord Pool and ECX, the last estimations for the Dec05 contract are from September to December of the respective year. For ECX, estimates for the third quarter 2007 (Q3 2007) are from June to September 2007. Results with less than 100 observations are not depicted.

Contract		ECX				Nord Pool				
		$\hat{s}/2$	in %	Adj. R^2	Obs.	$\hat{s}/2$	in %	Adj. R^2	Obs.	t-stat.
Dec05	2005	0.0624	0.27	0.21	2 256	0.0750	0.33	0.18	501	-0.98
	Q2	0.0614	0.31	0.29	445	0.0594	0.31	0.20	206	0.17
	Q3	0.0769	0.34	0.24	1 240	0.0933	0.40	0.18	270	-0.76
	Q4	0.0418	0.19	0.14	1 093	0.0657	0.29	0.24	156	-1.99**
Dec06	2006	0.0512	0.31	0.18	7 832	0.0890	0.54	0.08	1 140	-3.21***
	Q1	0.0448	0.17	0.24	2 032	0.0810	0.30	0.28	223	-3.34***
	Q2	0.0797	0.50	0.20	2 625	0.1252	0.79	0.07	503	-1.74*
	Q3	0.0362	0.22	0.25	1 493	0.0587	0.36	0.25	286	-3.20***
	Q4	0.0303	0.32	0.14	1 682	0.0347	0.32	0.16	128	-0.52
Dec07	2007	0.0178	1.42	0.21	2 412	0.0246	2.05	0.10	184	-1.06
	Q1	0.0213	1.02	0.23	1 636	0.0322	1.34	0.12	122	-1.30
	Q2	0.0118	2.19	0.22	572					
	Q3	0.0059	4.53	0.18	154					
Dec08	2007	0.0266	0.13	0.26	21 292	0.0563	0.26	0.25	216	-9.67***
	Q1	0.0404	0.27	0.30	3 333	0.0775	0.50	0.31	263	-4.05***
	Q2	0.0321	0.15	0.27	4 944	0.0696	0.31	0.28	722	-6.09***
	Q3	0.0216	0.11	0.27	7 303	0.0608	0.30	0.28	507	-6.86***
	Q4	0.0205	0.09	0.27	5 712	0.0315	0.14	0.20	724	-4.27***

decomposition of the spreads that are displayed in Table 2.4. We observe that for both exchanges the asymmetric information component $\hat{\theta}$ is significantly positive and constitutes by far the larger share of the traded spread. Note that the transitory component $\hat{\phi}$ is very small and sometimes even negative, especially for the Nord Pool contracts. Often it is not significantly different from zero.[21] One possibility to circumvent the negative sign of $\hat{\phi}$ is to set ρ and hence $E[x_t|x_{t-1}]$ equal to zero and thus ignore autocorrelation as other

---
[21]Note that estimating the model by OLS and accounting for serial correlation by the use of Newey-West standard errors, significance levels slightly increase.

trade indicator models do (see e.g. Glosten and Harris (1988)). Thereby, the magnitude of estimated half spreads does not change and we observe positive transitory components that are in most of the cases significant. However, the permanent component still accounts for the much larger share of the estimated half spread. Thus, both approaches yield qualitatively similar results such that the assumption about the conditional expectation does not alter our conclusions.

For all contracts there is a (local) peak in the permanent share in the second quarter of 2006. Hence, it appears that bid-ask spreads charged by liquidity providers in the European CO_2 market are mainly charged as a protection against losses to informed traders (see e.g. Bagehot (1971) or Glosten and Milgrom (1985)) and only to a marginal extent as a compensation for order processing or inventory costs. Given the extensive amount of uncertainty in the market about the development of price drivers as energy and fuel prices or about regulatory issues concerning future National Allocation Plans and the use of project based EUAs (CERs) as well as private news on the installation level, this result is not surprising.

As a last exercise, we assess the intraday pattern of estimated bid-ask spreads. Figure 2.4 plots intraday half spreads for both platforms again estimated over the full common sample period against trading hours. The intervals of the day that we use for estimations are from 08:00 to 09:59, from 10:00 to 11:59, from 12:00 to 13:59, and from 14:00 to 15:29 for ECX and Nord Pool. Since trading on ECX takes place until 18:00 CET, for ECX the last intraday estimates are from 15:30 to 18:00.[22] If the market processes information or resolves uncertainty during the trading day, we would expect to see spreads decline in the course of trading as observed in other markets.[23] Investigating the patterns in Figure 2.4, we observe that half spreads for the first two trading hours are always higher than for the last trading interval at the respective exchange. Considering e.g. the Dec06 contract, on average, the trading day on ECX (Nord Pool) starts with an estimated half spread of 0.06 Euro(0.10 Euro) and closes with a half spread of 0.05 Euro (0.07 Euro).

---

[22]Note that results are not affected by the choice of time intervals.
[23]See e.g. Madhavan et al. (1997) and the discussion in Biais et al. (2005)

2.4. SPREAD ANALYSIS

Figure 2.4: Intraday pattern of estimated half spreads

The figure displays intraday pattern of estimated half spreads on ECX and Nord Pool separated by the four contracts.

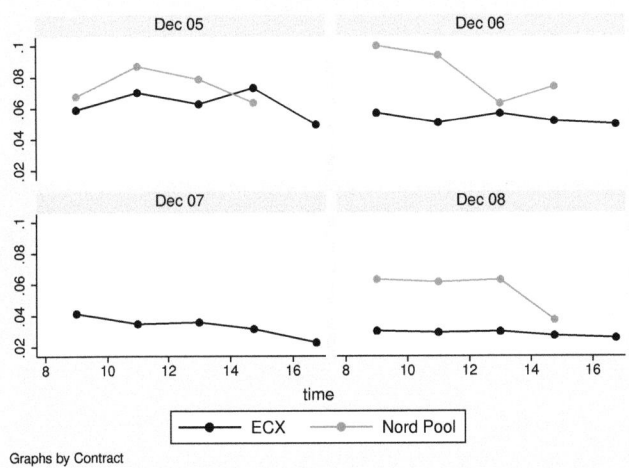

Regarding the share of the permanent spread component relative to the total spread, we find on ECX that except for the Dec07 contract, asymmetric information costs charged by liquidity providers decline over the course of the trading day.[24] For Nord Pool, however, no clear-cut picture emerges.

If the market development continues and overall uncertainty decreases in the future, we expect bid-ask spreads to decrease over the next trading phase 2008-2012 to levels close to 0.01 Euro. Overall, the development of transaction costs speaks in favor of a maturing market, in which traded volumes and trading intensity increase over time while transaction costs fall. Continuing from very low spread levels at the end of 2007, the EU ETS seems to be on a good way with respect to the functioning of organized EUA trading.

---

[24]Results are available from the authors upon request.

2.5 Price Discovery

In this section we want to determine which exchange is the first to process incoming information into prices. Note that the econometric model that we are going to apply in this section is not related to the bid-ask spread analysis from the previous section. However, the previous results may give some intuition for the expected outcomes with respect to price discovery.

Following common practice in the literature on financial and commodity markets, we approach the question of price discovery by specifying a vector error correction model (VECM). We proceed by describing the econometric methodology before applying it to the EUA futures market making use of high frequency data. As detailed above, observing lower transaction costs and higher trading volumes on the platform ECX, our prior is to expect a leading role for ECX. Correspondingly, before specifying an ECM, we use a conservative matching method that disfavors ECX, which we labeled earlier as NP-Match, in order to be sure that results are not driven by the use of newer ECX prices compared to Nord Pool.

Generally, the use of high frequency data is only more appropriate compared to daily data if events in the market under examination are also of high frequency. To obtain an intuition on whether this is the case, we compute the fraction of zero returns from one matched transaction in one market to the next one. For the matching approach favoring Nord Pool (NP-match), the fraction of zero returns for Nord Pool ranges from 0.19 to 0.34 for the different contracts. For ECX values are higher and range from 0.38 to 0.45. Compared to more mature markets, these figures are rather high. However, apparently there is more information in intraday data compared to daily data such that there is reason to use data of highest frequency.[25]

---

[25]Considering e.g. our third matching approach as suggested by Harris et al. (1995), figures are lower and range from 0.15 to 0.29 for Nord Pool and from 0.10 to 0.18 for ECX. These magnitudes can also be found on more mature financial markets.

2.5.1 Methodology

Cointegration and Error Correction

Generally, price discovery is the process by which markets attempt to find (discover) equilibrium prices by incorporating new information.[26] In case that an identical asset is traded at the same time in several markets, due to no-arbitrage arguments there should be no significant price differences across the markets. Formally, this means that there is an equilibrium price of the asset, which is common to all markets, and the sources of its price variation are attributed to different markets. While market efficiency implies that new information is impounded instantaneously into prices, markets process and interpret news at different rates (e.g. due to institutional factors such as transaction costs) and thus disequilibria occur, especially in an immature market like the EU ETS.

Explicitly, in our case of two markets this means that the two prices may be driven in a fundamental sense by one market, which is the price leader whereas the other market acts as a price taker. The price leader thus incorporates news faster into prices than the other market. Hence, returns on this market should lead the returns on the other market.

To investigate leadership in the EUA futures market, we apply two relative measures of price discovery which both use the VECM as their basis. Hence, we elaborate on the procedure by Engle and Granger (1987) showing that a VECM framework is appropriate for cointegrated time series.[27] The idea behind cointegration is that while two (or more) time series are non-stationary I(1) processes, they do not drift too far away from each other, such that their difference will be stationary. In that case, a proportion of the deviation from the equilibrium path in one period is corrected in the next period – the EC mechanism.[28]

---

[26] It has been argued that the process of price discovery in security markets is one of the most important products of a security market (cf. Hasbrouck (1995), p. 1175.)

[27] Note this approach is equivalent to estimating a VAR model of log returns on both exchanges augmented by a so called error correction (EC) term, i.e. a term including the difference between lagged prices on both exchanges.

[28] The relationship between ECM and cointegration was first pointed out in Granger (1981). A theorem showing precisely that cointegrated series can be represented by an ECM was originally stated and proved by Granger (1983).

Thus formally, returns should be represented by a VECM of the form

$$\Delta p_t = \mu + \sum_{k=1}^{K} \Gamma_k \Delta p_{t-k} + \alpha \beta' p_{t-1} + \epsilon_t \qquad (2.4)$$

where $p_t = (p_t^{ECX}, p_t^{NP})$ are the log futures prices on ECX and Nord Pool, μ and α are (2×1) vectors of parameters, Γ_k are (2×2) matrices of parameters, and K is the lag-length, which will be determined by the Schwarz criterion. ϵ_t is a (2×1) error vector with mean zero and variance-covariance matrix Ω, Δ is the difference operator (e.g. $\Delta p_t = p_t - p_{t-1}$) and β is the (2×1) cointegrating vector, which is in our case equal to (1 -1)'.[29] In this model the current returns are then explained by (i) the past returns on both markets (short-run dynamics induced by market imperfections), and (ii) the deviation from the no-arbitrage equilibrium (long-run dynamics between the price series), i.e. $p_t^{ECX} - p_t^{NP}$.[30] Consequently, the cointegrating vector defines the long-run equilibrium, while the EC dynamics characterize the price discovery process. Note that the coefficient vector of the EC term $\delta = (\delta^{ECX} \; \delta^{NP})$ is (by construction) orthogonal to the EC coefficient vector α. This coefficient vector is needed to compute the following two common factor measures for price discovery.

Common Factor Measures

One measure has been introduced by Schwarz and Szakmary (1994) and only regards the EC process, i.e. only δ is relevant. The other measure has been suggested by Hasbrouck (1995) and additionally takes into account the variance of the innovations to the common factors of the price series.[31]

---

[29] We did not explicitly estimate but rather pre-specified the cointegration vector since the long-run equilibrium is given by $p_t^{ECX} - p_t^{NP} = 0$, see e.g. Theissen (2002). This relation holds for our case as we are using discounted price series.

[30] Compare also Baillie, Booth, Tse and Zabotina (2002), p. 311.

[31] For a comparison of both measures, see the discussion by De Jong (2002); Baillie et al. (2002).

2.5. PRICE DISCOVERY

Common Factor Weights (CFW)

Schwarz and Szakmary (1994) argue that the coefficients δ^{ECX} and δ^{NP} in the VECM in Equation (2.4) represent the permanent effect that a shock to one of the variables has on the system. Therefore they propose to use the relative magnitude of these coefficients to assess the contributions of the two trading systems to price discovery.[32] Specifically, they propose the measure

$$CFW^{ECX} = \frac{\delta^{NP}}{\delta^{NP} - \delta^{ECX}}; \quad CFW^{NP} = \frac{-\delta^{ECX}}{\delta^{NP} - \delta^{ECX}}. \tag{2.5}$$

A high magnitude of δ_i (i = ECX, NP) in the respective market corresponds to slow information dissemination. Apart from describing adjustment dynamics, the coefficients measure the speed of assimilation to discrepancies between the markets. Thus, the common factor weights quantify the share of total reaction being attributable to one market.

Information Shares (IS)

The information share approach of Hasbrouck (1995) relates the contribution of an individual market's innovation to the total innovation of the common efficient price instead of only focusing on coefficients of the deviation term. To derive the IS formula, Hasbrouck transforms Equation (2.4) into a vector moving average (VMA)

$$\Delta p_t = \Psi(L)e_t. \tag{2.6}$$

Its integrated form can be written as

$$p_t = p_0 + \Psi(1)\sum_{s=1}^{t} e_s + \Psi^*(L)e_t, \tag{2.7}$$

---

[32] A formal justification can be derived from the work of Gonzalo and Granger (1995).

where p_0 is a vector of constant initial values, $\Psi(L)$ and $\Psi^*(L)$ are matrix polynomials in the lag operator, L, and the (2×2) matrix $\Psi(1)$ is the sum of the moving average coefficients. It is called the impact matrix as $\Psi(1)e_s$ (for $s = 1,...,t$) measures the long-run impact of an innovation on each of the prices. Due to the pre-specified cointegration vector $\beta = (1\ -1)'$ the long-run impact is the same for both prices. This translates into an impact matrix whose rows are identical. With $\psi = (\psi_1\ \psi_2)$ being the common (1×2) row vector of $\Psi(1)$, Equation (2.7) becomes

$$p_t = p_0 + \iota(\psi \sum_{s=1}^{t} e_s) + \Psi^*(L)e_t, \qquad (2.8)$$

where $\iota = (1\ 1)'$ is a column vector of ones. While $\Psi^*(L)e_t$ simply denotes the transitory portion of the price change, Hasbrouck defines the first part of Equation (2.8) – the random-walk component – as the common factor component or the common efficient price in the two markets.[33] The common factor innovations (increments) ψe_t (for $s = 1,...,t$) are the components of the price change that are permanently impounded into the price and that are presumably due to new information. Thus, we are interested in this part when analyzing the process of price discovery.

We observe that the innovations' covariance matrix Ω is not diagonal as price innovations are correlated across the two markets. To investigate the proportion of the total variance in the common efficient price that is attributable to innovations in one of the two markets (hence, its information share), the variance of the common factor innovations $\text{Var}(\psi e_t) = \psi\Omega\psi'$ has to be decomposed. The Cholesky factorization of $\Omega = MM'$ can be applied to minimize contemporaneous correlation, where M is a lower triangular (2×2) matrix.[34]

---

[33] His specification is closely related to the common trend representation of prices from different markets in Stock and Watson (1988).

[34] Hasbrouck (1995) states that most of the contemporaneous correlation comes from time aggregation as in practice, market prices usually change sequentially. As one way to minimize the correlation, he suggests to shorten the interval of observation and to synchronize the data. However, as this will only lessen but not eliminate the contemporaneous correlation, he additionally proposes the triangularization of the covariance matrix.

2.5. PRICE DISCOVERY

The information shares are given as follows:

$$S_j = \frac{([\psi M]_j)^2}{\psi \Omega \psi'}. \tag{2.9}$$

There are many different factorizations of Ω. Due to the nature of the Cholesky decomposition, the lower triangular factorization maximizes the information share of the first market and consequently minimizes the share of the second market. Thus, by permuting ψ and Ω, upper and lower bounds for each market's information share are obtained. Following the literature, we use the mean of the upper and the lower bound as a unique measure of a market's information share. As formally justified by e.g. Martens (1998) and Theissen (2002), the common row vector ψ is directly related to the coefficient vector of the EC term δ, i.e. $\frac{\psi_1}{\psi_2} = \frac{\delta_1}{\delta_2}$. Together with Equation (2.9) and by noting that $S_{ECX} + S_{NP} = 1$ the IS can be rewritten as

$$S_1 = \frac{(\delta_1 m_{11} + \delta_2 m_{21})^2}{(\delta_1 m_{11} + \delta_2 m_{21})^2 + (\delta_2 m_{22}^2)} \tag{2.10}$$

$$S_2 = \frac{(\delta_2 m_{22})^2}{(\delta_1 m_{11} + \delta_2 m_{21})^2 + (\delta_2 m_{22}^2)}. \tag{2.11}$$

Both equations show that the IS only depend on the vector α (or its orthogonal vector δ) and Ω.[35] They also show that the factorization imposes a greater IS on the price of the first market (unless $m_{21} = 0$, i.e. no correlation between market innovations exists).

2.5.2 Estimation Results

After applying stationarity and cointegration tests to our price series, we present the estimation results of price discovery in two parts. The first part describes the results of the VECM estimations, the second part the information shares and the common factor weights. Note that we applied the methodology to each data synchronization scheme. As

---

[35]Note that we use a different matrix indexing than Baillie et al. (2002). With m_{21} we denote the entry of the second line in the first column of the matrix M.

the NP-Match is of highest interest for our study, we explicitly report its estimation results and only verbally describe deviations from the other two matches.

Stationarity and Cointegration Tests

Proper interpretation of cointegration models requires that all futures prices contain a single unit root implying non-stationarity. To test for stationarity we apply the well-known Augmented Dickey-Fuller (ADF) unit root test as well as the Kwiatkowski, Phillips, Schmidt, and Shin (KPSS) stationarity test.[36] For both tests the truncation parameter to select the autocorrelation lag length is chosen according to the Schwarz information criterion. Table 2.5 presents the results of the unit root tests for the whole sample period for the NP-Match (April 2005 to December 2007).[37] For the Dec05 and Dec08 contract we do not explicitly consider a trend in the unit root test as a visual inspection of the data fails to provide an indication of a trend (see e.g. Uhrig-Homburg and Wagner (2007)).

With respect to the log price series, ADF tests reject the null hypothesis of a unit root only for the Dec08 contract on both exchanges. The KPSS tests reject the assumption of stationarity for all contracts except for Dec06. We conclude that the evidence is in favor of non-stationarity as indicated by the mostly insignificant ADF and significant KPSS tests, respectively. For the first-differenced series both tests are almost completely in favor of stationarity, only on ECX stationarity is rejected at the 10% level for the Dec05 contract. Note that applying the unit root tests to the ECX-Match we get the same picture as for the NP-Match. For the Harris-Match, the tests clearly indicate non-stationarity of the log prices and of stationarity for the first differences. Testing for cointegration, we use the likelihood ratio test procedure proposed by Johansen (1988, 1991). The results indicate that the time series from Nord Pool and ECX are cointegrated.

---

[36]While the null hypothesis for the ADF test is the existence of a unit root, the KPSS test assumes a stationary time series.

[37]Note that futures trading on Nord Pool started in February 2005, while ECX trading was launched in April 2005.

2.5. PRICE DISCOVERY

Table 2.5: Results for stationarity tests

The table presents the test statistics from Augmented Dickey Fuller (ADF) tests and Kwiatkowski-Phillips-Schmidt-Shin tests (KPSS) applied to both price levels and the first differences of the time series on ECX (upper panel) and Nord Pool (lower panel).*,**, and *** stand for rejection at the 10, 5, and 1 per cent levels.

ECX	level		first difference	
	ADF	KPSS	ADF	KPSS
log(Dec05)	-2.058	0.632**	-25.371***	0.365*
log(Dec06)	-2.444	0.269	-35.306***	0.068
log(Dec07)	-1.624	0.318*	-22.028***	0.076
log(Dec08)	-3.340***	3.67***	-66.800***	0.224

NP	level		first difference	
	ADF	KPSS	ADF	KPSS
log(Dec05)	-2.013	0.648**	-25.403***	0.299
log(Dec06)	-2.567	0.269	-36.878***	0.068
log(Dec07)	-1.613	0.339*	-21.718***	0.068
log(Dec08)	-3.536***	3.690***	-69.522***	0.225

Error Correction Model

We apply the VECM derived above to the synchronized high frequency EUA futures log price series. The VECM in Equation (2.4) can be written as

$$\Delta p_t^{ECX} = \alpha^{ECX} + \sum_{k=1}^{K} \gamma_{11,k} \Delta p_{t-k}^{ECX} + \sum_{k=1}^{K} \gamma_{12,k} \Delta p_{t-k}^{NP} + \delta^{ECX}(p_{t-1}^{ECX} - p_{t-1}^{NP}) + \epsilon_t^{ECX} \quad (2.12)$$

$$\Delta p_t^{NP} = \alpha^{NP} + \sum_{k}^{K} \gamma_{21,k} \Delta p_{t-k}^{ECX} + \sum_{k}^{K} \gamma_{22,k} \Delta p_{t-k}^{NP} + \delta^{NP}(p_{t-1}^{ECX} - p_{t-1}^{NP}) + \epsilon_t^{NP}.$$

The coefficients δ^{ECX} and δ^{NP} determine the speed of adjustment of the respective price towards the long-run equilibrium levels, which is assured by the no-arbitrage argument. If

Table 2.6: Estimation results of the VECM for Phase 1

The table presents the CFW for both markets and for all contracts together with the information shares for ECX. We report the mean of the upper and lower bound and the corresponding range (difference between upper and lower bound). A '++' or '+' indicates that the coefficients of the error correction vector ($\hat{\delta}^{ECX}, \hat{\delta}^{NP}$) are significantly different from 0 at the 5% or 10% level, respectively. For the Dec08 contract, we only use transaction prices from the year 2007.

Contract	EC		CFW		IS for ECX		Obs.
	δ^{ECX}	δ^{NP}	ECX	NP	Mean	Range	
Dec05	++	++	0.593	0.407	0.546	0.790	615
Dec06	+	++	0.811	0.189	0.623	0.725	1433
Dec07		++	0.830	0.170	0.714	0.513	413
Dec08		++	0.847	0.153	0.644	0.693	2402

ECX incorporates information faster, we expect δ^{ECX} to be insignificant, while δ^{NP} should be significant and bear a positive sign.

Table 2.6 presents estimated common factor measures of the VECM estimation of the NP-Match for all four futures contracts and covers the common sample period.[38] To conserve space, we do not display the coefficients of the EC term ($\delta^{ECX}, \delta^{NP}$) and of the VAR terms $\gamma_{ij,k}$, and only report the CFW for both markets. The coefficients' level of significance is marked by '++' or '+', which indicates that they are significantly different from 0 at the 5% or 10% level, respectively. Furthermore, the table depicts information shares for ECX. It reports the mean of the upper and lower bound and the range (difference between upper and lower bound). Remember that upper and lower bounds are obtained from changing the ordering in the Cholesky factorization. For the Dec05 contract we include 2 lags, and for Dec06/ Dec07/ Dec08 we take 8/ 1/ 3 lags, respectively.[39]

We find that for all contracts in both equations of (2.12) the coefficient of the EC term has the expected sign and is significant at least in one of the markets. Thus, price discovery takes place. Apparently, for Dec05 and Dec06 both markets contribute to the process of

---

[38] As for Dec08 trading activity was very low in 2006, we start the analysis in January 2007.
[39] We applied the Schwarz information criterion for the whole sample period as well as for three-month intervals displayed in Table 2.7. As final lag-length we took the maximum.

2.5. PRICE DISCOVERY

price discovery. However, ECX is the clear price leader for the Dec07 and Dec08 contracts. Measuring the markets' contribution to price discovery both measures tend in the same direction. We find that for later expiration dates price discovery increasingly takes place on ECX. These results are in line with the development of the EUA futures market. As stated in the introduction, Nord Pool was the first platform which started to trade EUA futures and is thus expected to be the more experienced market for the first months of trading. ECX joined some time later and managed to attract more liquidity in the course of the time.

To counteract critique of analyzing a too long data sample and to account for structural breaks, we zoom into the most liquid trading phase of each contract and divide it into calendar quarters. Note that due to the lack of observations we start the analysis for the Dec05 contract with the second quarter and we remove the Dec07 contract from the quarterly analysis.[40] The results in Table 2.7 reveal the following interesting pattern for price discovery: Both measures indicate that ECX's contribution peaks in the second (Q2) and third quarter (Q3) compared to the first (Q1) and last (Q4) quarter. An exception constitutes the Dec05 contract, where in the second quarter price discovery still takes place on both platforms. To find possible explanations for this behavior, we analyze quarterly trading activity measured by the average number of daily transaction frequencies and average daily trading volume. Apparently, Table 2.7 states that the observed price discovery pattern is mostly in line with the one for trading activity: whenever liquidity is increasing ECX mostly leads the price whereas Nord Pool's contribution becomes again observable when transaction frequencies and trading volumes decline. An exception is the sharp decrease in liquidity after the second quarter for the Dec06 contract, which did not lead to a change in the CFW. As was stated, this behavior reflects the announcement of an considerable over-allocation of EUAs at the end of April, which led to a substantial drop in demand for EUAs and thus to a drop in spot and futures prices. Furthermore, it might be the case that findings for the last quarter are related to an earlier expiry of Nord Pool

---

[40]In case that there are less than 190 observations, the preceding month is included into the analysis as indicated at the bottom of the table.

Table 2.7: Estimation results of the VECM for a restricted period, daily transaction frequencies, and trading volume

The table presents the CFW for both markets and for all contracts together with the information shares for ECX for the quarters of the most liquid trading year. We report the mean of the upper and lower bound and the corresponding range (difference between upper and lower bound). A '++' or '+' indicates that the coefficients on the error correction vector ($\hat{\delta}^{ECX}, \hat{\delta}^{NP}$) are significantly different from 0 at the 5% or 10% level, respectively. Furthermore, daily transaction frequencies (TA), and the trading volume are presented. In Q4 for the Dec05 and Dec06 contracts the September is included as there are less than 150 observations.

Contract		EC		CFW		IS for ECX		TA		Volume		Obs.
		δ^{ECX}	δ^{NP}	ECX	NP	Mean	Range	ECX	NP	ECX	NP	
Dec05	Q2 2005	+	+	0.513	0.487	0.513	0.824	11	4	79.8	34.5	179
	Q3		++	0.611	0.389	0.563	0.744	23	5	171.7	42.4	326
	Q4	++		0.133	0.867	0.452	0.897	17	3	141.4	25.8	199
Dec06	Q1 2006			0.565	0.435	0.508	0.940	41	5	415.2	26.2	269
	Q2		++	0.785	0.215	0.624	0.706	57	9	517.2	68.3	540
	Q3		++	0.794	0.206	0.626	0.712	30	6	248.3	41.3	340
	Q4		+	0.688	0.312	0.547	0.866	47	5	313.0	34.0	278
Dec08	Q1 2007			0.367	0.633	0.485	0.932	77	6	675.2	38.1	307
	Q2		++	0.920	0.080	0.765	0.461	145	13	966.7	78.5	775
	Q3		++	0.929	0.071	0.734	0.526	178	10	1296.7	71.3	556
	Q4	+	++	0.683	0.317	0.555	0.831	177	13	1087.4	98.1	764

contracts compared to ECX futures.

When interpreting our results it should be kept in mind that the construction of our dataset, NP-Match, puts ECX at a disadvantage and thus favors the less liquid market Nord Pool. Hence our results are likely to even understate the role of ECX in the process of price discovery. To check the robustness of our results we estimate the VECM of Equation (2.12) also for the ECX- and Harris-Match. While results from both matches are even more in favor of ECX they also show that Nord Pool significantly contributes to price discovery in the first and last quarters of the most active trading year. We hence conclude that while ECX is the clear price leader in the EUA futures market, our null hypothesis of no contribution to price discovery by Nord Pool has to be rejected.

2.6 Conclusion

In our chapter we analyzed high frequency data for European Union Emissions Allowance (EUA) futures for the whole first trading period. Data has been provided by the two most liquid trading platforms ECX and Nord Pool. After having given a short market overview we addressed the issue of market liquidity. We conducted a spread analysis by applying a trade-indicator model. Having two cointegrated price series we were able to measure the process of price discovery by estimating a vector error correction model.

Our results revealed that estimated transaction spreads markedly decreased on both exchanges over time and were lower on ECX than on Nord Pool. With respect to price discovery, we demonstrated that for the first EUA futures contracts, Dec05 and Dec06, both exchanges contributed to price discovery. However, for the most recent contracts, Dec07 and Dec08, the more liquid market ECX became the price leader, especially in phases of high market liquidity but Nord Pool's contribution was still present in times of lower transaction frequencies and volumes.

Obviously, our results are not only of academic interest. They indirectly give several market recommendations. First and most obviously, in order to remain (as second competitive platform) in the market and not to lose further market share to ECX, Nord Pool should take some action to attract liquidity. The same is true for other existing market competitors, especially given the large (but decreasing) extent of competition from the OTC market. Besides, the sharp increase in trading volumes over time in this very young market reveals that there may be a lot of profits for other trading platforms and market participants from entering the (futures) market.

The development potential of the EUA market is extremely high since the EUA can be considered as an European and – depending on future regulatory decisions with respect to additional member states – as a global asset. Low correlations with other financial assets and commodities together with an increasing range of derivative products have furthermore increased the attractiveness of EUAs as an asset class. Hence we would expect to see rapidly increasing interest from the banking as well as the mutual and hedge fund industries in the market such that in the future, compliance trading may no longer constitute the largest share of EUA trading.

Summing up, together with a rapid expansion of the market for EUAs and CERs, in the near future we expect to observe an increasing number of platforms that try to participate in the growing and promising market before seeing a phase of consolidation after which some main trading platforms will emerge.

Recent developments in the carbon market support these statements. In December 2007 Nord Pool has announced to merge with the Nordic exchange OMX to attract further liquidity. Besides, EEX started a cooperation with EUREX in order to increase their market share in EUA futures trading for the Phase 2 and beyond. Not only already established platforms aimed to expand, also new market platforms decided to join the market. In spring 2008 the US American Green Exchange, a cooperation of NYMEX and the environmental broker EvolutionMarkets, launched EUA and CER futures contracts for the years 2008 to 2012. BlueNext, a cooperation between NYSE Euronext and Caisse des Depots was formed in December 2007. It only specializes in carbon related products that have been acquired from Powernext.

Thus, carbon indeed becomes an internationally traded commodity and there is awareness of this steadily growing market. Consequently, the importance of a well functioning market and the guarantee of smooth trading systems are essential. The instruments we are using in our analysis give evidence that after having some difficulties at the beginning, the carbon market is now able to fulfill these requirements. It is possible to track the process of price discovery with the development of the market and to identify the market platform, which is informationally dominant. Besides, bid-ask spreads can be used as a benchmark for liquidity. As our study is the first that includes the additional trading year 2007 in which liquidity has increased significantly and that provides intraday transaction prices, we have the advantage to obtain a more detailed insight into trading patterns compared to prior studies, which we can use and to investigate efficiency measures.

We conclude that as the design for EUA market platforms seems to work and as at least some form of "operational efficiency" has been achieved in the market, the regulatory

2.6. CONCLUSION

authorities can concentrate more on issues like the initial allocation process for the EU ETS that have not yet been solved for the upcoming post Kyoto trading period.

Chapter 3

The Initial Allocation of CO_2 Emission Allowances: An Experimental Study

> This chapter seeks an appropriately designed initial allocation mechanism of CO_2 allowances for the third trading period (2013-2020) of the European greenhouse gas emissions trading scheme. We consider grandfathering, auctions, and combinations thereof and postulate three main criteria for a viable initial allocation mechanism: information efficiency, allocation efficiency, and incentive compatibility. Keeping these criteria in mind, we analyze and evaluate four policy-relevant allocation rules both theoretically and experimentally. Assuming that participating firms engage in myopic bidding behavior, we demonstrate that only two initial allocation rules fulfill these criteria. For the two allocation rules that employ grandfathering, a uniform double auction proves to be superior to a uniform one-sided auction. However, for a rule based solely on auctioning, a uniform one-sided auction is also an appropriate mechanism, provided that firms do not possess any allowances at the time of the auction.

3.1 Introduction

The EU-wide greenhouse gas emissions trading scheme (EU ETS) was formally launched in January 2005. Affecting about 50% of Europe's CO_2 emissions and 40% of its total greenhouse gas emissions, the scheme is predicated upon a finite number of tradable emissions allowances that are allocated among firms on an annual basis. One allowance entitles the holder to emit one ton of CO_2 greenhouse gases. Firms in possession of more allowances than they need to cover their emissions may sell them to firms that do not have enough.

This mechanism gives firms an incentive to abate emissions where it is cheapest to do so and thereby the required level of emissions can be reached at the lowest cost.

The scheme consists of trading phases and entails an annual allocation of allowances to participating firms for the current trading year. Implementing a politically viable initial allocation rule constitutes a major challenge for the scheme's regulator, which is the European Commission. Currently, the Commissions is concerned with creating an initial allocation rule for the third trading period of the EU ETS, also known as Phase 3, which will run from 2013 to 2020.

Our objective is to recommend a viable initial allocation process for Phase 3. To this end, we examine and evaluate several policy-relevant initial allocation mechanisms for the EU ETS and rate them according to their conformance to three main criteria: Information efficiency, i.e. the generation of early and reliable allowance price signals, helps firms to make correct abatement decisions. Allocation efficiency entails that the firms with the greatest need, i.e. those with the highest abatement costs, receive the allowances. Incentive compatibility means that firms need only consider their own abatement costs when formulating their individual bidding strategies. Although additional criteria like revenue maximization, political feasibility, and social acceptance also have to be considered, they are secondary from the firms' point of view and thus do not constitute the focus of our analysis. Using both theoretical and experimental frameworks, we compare four different allocation rules in combination with a secondary market. Our experimental analysis reveals that subjects tend to behave myopically when deciding on buying or selling allowances in that they only take their own abatement costs into consideration. Under this bidding behavior, two of the four allocation rules under investigation fulfill all three criteria: grandfathering (gratis allocation) plus a double uniform auction (in which firms may act as both buyers and sellers) and an exclusive one-sided uniform auction (buying position only). The other two allocation rules, i.e. grandfathering plus a one-sided uniform auction, which is the top contender for Phase 3, and exclusive grandfathering, which was mostly used for Phase 1 and continued for Phase 2 of the EU ETS, fail to conform to the criteria. Note that the mechanism of a double auction, whose total auction supply is the sum of the supply issued

3.1. INTRODUCTION

by the government and the firms, is principally equivalent to a continuous trading scheme, in which the governmental supply is offered on trading platforms for allowances at regular and publicly announced time intervals. The German government uses this modified form of auctioning in order to auction 8.8% of the total national allowance budget for the years 2008 and 2009 of Phase 2 (Zuteilungsgesetz, 2007).

Conducting a laboratory experiment is a powerful tool to study the economic effects of different initial allocation mechanisms for CO_2 allowances that focus on auctioning. The most important advantage is that institutions for emission allocation are exogenously changed. Thus, it is possible to observe the ET market before and then after auctions are introduced into the allocation process, holding everything else constant. This methodology allows controlling for crucial aspects of the economic environment, such as information conditions, market structures or economic trends. This helps to disentangle the impact of a change in institutions. Besides, control over the decisions environment makes it possible to identify the theoretically optimal level of abatement (theoretical optimum) in an experimental ET market and, given those theoretical predictions, to analyze the impact of institutions on micro behavior. Thus, with a laboratory experiment we can study initial allocation rules that do not yet exist, at relatively low costs.[1]

The remainder of the chapter paper is organized as follows. In Section 3.2 we introduce policy-relevant initial allocation rules and describe the three criteria that they are supposed to meet. Then, we briefly integrate our study into the existing (experimental) ET literature in Section 3.3. Section 3.4 describes the experimental setup and Section 3.5 formulates theoretical considerations towards the three criteria. They build the basis of the experimental analysis in Section 3.6, where results are analyzed, and discussed in detail in terms of compliance with the criteria. The paper ends with a brief conclusion in Section 3.7.

---

[1] For experimental literature with respect to other market institutions than ET, such as labor or financial markets, see e.g. Falk and Huffman (2007) or Ackert and Church (2001), Ehrhart (2001), respectively. Holt (1989) for example studies market power in laboratory market institutions.

3.2 Initial Allocation Rules and Criteria

At the end of every February, a certain amount of allowances, i.e. the initial allocation, is allocated to participating firms for the current trading year according to so-called "National Allocation Plans". Typically two types of allocation rules are employed, auctioning and grandfathering (gratis allocation in proportion to historical emissions), either alone (exclusive scenarios) or in combination (hybrid scenarios). The EU ETS consists of consecutive trading periods, Phase 1 (2005-2007), Phase 2 (2008-2012), and Phase 3 (2013-2020). The initial allocation rules for the first two phases are stipulated in the "ETS Directive". Accordingly, EU Member States were permitted to auction off up to 5% of their total allowances in Phase 1 and up to 10% in Phase 2. Four Member States permitted auctioning in Phase 1, and eight countries are allowing auctioning in Phase 2; in any case, the shares to be auctioned are well below the allowed maximum shares. A recent proposal for a new Directive by the European Commission for Phase 3 implies that about two thirds of the total allowances will be auctioned off initially. Explicitly, operators of energy installations will have to purchase all of their allowances, whereas industrial installations will probably start with an initial auctioning share of 20% in 2013, to be increased to 100% by 2020.

With respect to the auction design itself, it is very likely that authorities will decide in favor of a one-sided auction mechanism because it is both the most common and easiest format for auctioning homogenous goods. For instance, in Phase 1 of the EU ETS, Hungary and Ireland conducted static one-sided auctions; meanwhile, dynamic one-sided auctions have already been applied by the US market for NO_x and by the UK ETS for CO_2 allowances. However, in all these markets, the sole reason for auctioning was to raise revenue to finance administrative costs (Evans and Peck, 2007). Furthermore, Holt, Shobe, Burtraw, Palmer and Goeree (2007) propose a one-sided auction format to allocate allowances for the Regional Greenhouse Gas Initiative (RGGI) in the US starting in 2009.[2] However, the US Acid Rain Program for SO_2 permits – the most prominent ETS before the EU

---

[2]RGGI is a cooperative effort by Northeastern and Mid-Atlantic states to reduce CO_2 emissions in the electricity sector by means of ET. This program has already agreed to auction at least 25% of the emission allowances in 2008.

3.2. INITIAL ALLOCATION RULES AND CRITERIA 83

ETS– conducts static double auctions in order to generate price signals and achieve an efficient allocation (Cason, 1993; Schmalensee, Joskow, Ellerman, Montero and Baily, 1998). Consequently, a closer look at the SO_2 auctions might be a good source of inspiration for developing an appropriate EU carbon auction design.

The initial allocation determines the firms' stock of allowances and consequently influences which activities the firms will undertake in order to comply with their individual CO_2 reduction targets. Therefore, a viable initial allocation rule should enable firms to abate emissions at different prices (costs); also, buying and selling allowances at competitive prices should be relatively easy and transaction costs ought to be low. The ideal allocation rule should therefore fulfill the following three criteria (C1 to C3).

(C1) Information efficiency

This criterion is associated with early and "reliable" price signals, which allow firms to profitably invest in cost-efficient abatement measures. A price signal is said to be reliable if it reflects the true scarcity of emission allowances in the system. Criterion C1 can only be satisfied by allocation rules that incorporate auctioning because auctions generate price signals whereas grandfathering does not. The criterion of correct price signals is essential for planning and realizing abatement projects. Most emissions-reducing investment projects, especially those with a high degree of energy efficiency, are costly and involve long implementation times. These projects thus require a long-term planning horizon. Therefore, the initial allocation mechanism should ideally facilitate and support the generation of early and reliable price signals.[3]

(C2) Allocation efficiency

This criterion stipulates that the firms with the highest willingness to pay (derived from their abatement costs) should receive the allowances. In other words, allowances are to be allocated to these firms above all others needing allowances to comply

---

[3] Apart from auctions, the prices generated on the futures markets for CO_2 emission allowances can also be considered as price signals. As auctions might be more relevant to smaller firms that do not have the capacity to speculate in the futures markets for CO_2 allowances, we study carbon auctions only.

with their commitment. Since the allocator is not aware of individual needs and grandfathering does not take them into account, Criterion C2 can only be satisfied by allocation rules that incorporate auctioning. In addition, an allocationally efficient auction reduces the necessity for allowance trading, resulting in lower trading costs, another criterion mentioned e.g. by Stavins (1994) and Cramton and Kerr (2002).[4] Ideally, no trading activities are necessary if auctions are implemented by the initial allocation mechanism only.[5] Therefore, the initial allocation mechanism should be designed to support the generation of an allocationally efficient outcome, which will automatically lowers trading costs.

(C3) Incentive compatibility

This criterion entails that the initial allocation mechanism should ultimately strive to be incentive compatible, i.e. a firm ought to be able to arrange its buying and selling strategies strictly according to its marginal abatement costs (MAC), which describe the firm's private valuation for one emission allowance. We call this bidding behavior "straightforward" or "myopic". Firms that engage in this type of behavior have no incentive to change their bidding strategy, because they observe the same price level during the auction and trading process. This additional criterion, which we call "stability", holds automatically when C3 is fulfilled. Obviously, it applies to auctions only. As a criterion, incentive compatibility is basic and desirable, especially when one considers that for the majority of firms, participating in the ETS plays a relatively minor role in their daily business. Thus, successful participation in the ETS should be possible without any outside knowledge (e.g. about other firms' characteristics or total scarcity conditions).

---

[4]The authors argue that auctions pose an attractive allowance allocation mechanism in the presence of trading costs, as regular participation in the secondary market incurs higher transaction costs for the firms than sporadic auction participation.

[5]Moreover, there is evidence that markets are not perfect and that a "helping hand model" is needed to allocate scarce resources. Ehrhart, Hoppe and Schleich (2006) show that the degree of cost-efficiency in ETS depends on the initial allocation: the more allowances are initially allocated to the firms that need them most (i.e. the firms with the highest abatement costs), the higher the degree of efficiency will be.

3.3 Related Literature

Though the experimental economics literature about ETS is comprehensive, the initial allocation rule has not been analyzed extensively as an experimental treatment variable. To our knowledge, there are no experimental studies that address the EU ETS. Early studies observed the performance of different trading institutions compared to command-and-control instruments in order to determine whether tradable emission schemes should be implemented or not. In these studies, however, the initial allocation rule is always treated as a given parameter and never as the object of investigation (e.g., Klaassen, Nentjes and Smith (2001); Mestelman, Moir and Muller (1999); Muller and Mestelman (1994, 1998); Plott (1983)). Later studies concentrated on the implementation of trading schemes with respect to the initial distribution of allowances among participants when grandfathering is applied (Ehrhart et al., 2006) or considered the implications of a ban on the banking of allowances (e.g., Godby, Mestelman, Muller and Welland (1997); Cronshaw and Brown-Kruse (1999); Cason, Elliott and Van Boening (1999)). Except for Ehrhart et al. (2006), none of the studies captures the unique institutional design of the EU ETS.

Of the literature on the SO_2 trading scheme, the work by Cason (1993, 1995) is very useful. He studies the allocation process of SO_2 permits via annual sealed bid/sealed offer discriminatory auctions with the auction rule of a "low-offer-to-high-bid" matching system. By means of a theoretical model and an experiment, he demonstrates that the implemented discriminatory price rule induces sellers and buyers to misrepresent their true values of the emission permits (cost for emission control) and to state lower asking and bid prices, because this increases their trading priority.[6] Conducting an experiment for testing the EPA auction with uniform pricing, Cason and Plott (1996) obtain a higher efficiency level, a more truthful revelation of underlying values and costs, and thus more accurate price information. The most recent auction experiments come from Porter, Rassenti, Shobe, Smith and Winn (2006) for the NO_x auctions in Virginia and from Holt et al. (2007) for the

---

[6] The lower the stated bid, the less likely it is that any other seller has a lower bid, which increases the probability of winning, i.e. sellers' bids only determine their probability of winning.

US RGGI. They test several one-sided auction setups. They only analyze the mechanisms with respect to revenue raising and partially measure allocation efficiency. However, in the aforementioned experimental literature, which addresses allocation rules for ETS, subjects do not acquire allowances to produce or to satisfy an exogenously imposed compliance cap. The authors employ a simplified, abstract commodity trading environment and ignore the opportunity of a resale market after the initial allocation process and the relevance of the existence of an initial allowance endowment (via grandfathering or banking) when auctions are involved. Furthermore, they do not address the importance of price signals generated by the different auctions and overlook the importance of an incentive compatible mechanism.

With our analysis we fill the gab in the literature by studying CO_2 initial allocation mechanisms that are embedded in a complete trading system. This holistic approach enables us to account for the temporal interaction of all of the system's components: initial allocation, trading, and abatement decision. Theoretically and experimentally, we provide a situation in which subjects act as profit-maximizing firms that have to decide on strategies in a stylized trading environment modeled on the EU ETS.

3.4 Experimental Design

Our experimental design embodies the main features of the EU ETS with some simplifications to prevent the system from becoming too complex to be controlled. As stated in the introduction, we analyze a system of initial allocation and trading.[7] We conduct four different variations of the initial allocation rule. All variants are based on the same ET game, which is described below. The four treatments are presented afterwards.

---

[7]This reflects the situation of Phase 3 in the EU ETS but also other future CO_2 markets from countries that have signed the Kyoto Protocol but so far do not participate in CO_2 ET (e.g., Australia, Canada, New Zealand) as well as in voluntary markets that have been established in e.g. the US, such as the RGGI.

3.4. EXPERIMENTAL DESIGN

3.4.1 Emissions Trading Game

One period in our ET game consists of two stages. In stage 1 subjects can buy and sell CO_2 allowances in an auction with an exogenous auction supply and in stage 2 subjects can trade CO_2 allowances on the market. Each subject represents a committed firm that has constant business-as-usual emissions of CO_2 in every period. It is the task of each firm to cover each ton of emissions with the same amount of CO_2 allowances. To do so, each firm has three possibilities: participating in stage 1 and/or stage 2 and/or switching to a more energy efficient technology to abate CO_2 emissions (abatement measure). Each firm has one abatement measure that is characterized by a maximum abatement volume of CO_2 and constant marginal abatement costs (MAC). Firms differ with respect to their MAC such that there are low-, middle-, and high-cost abatement technologies. To be able to participate in the auction and trading processes at the beginning of the period, each subject is given an individual monetary endowment measured in Experimental Currency Units (ExCU). Additionally, when grandfathering is part of the initial allocation rule, firms possess an initial stock of CO_2 allowances.

Note that if by the end of the period a firm lacks sufficient allowances to meet its CO_2 commitment (level of business-as-usual emissions), the firm's abatement measure is automatically implemented to cover any remaining emissions. The abatement costs per lacking ton are given by the subject's individual MAC. We decided in favor of automatic abatement in order to simplify the participation process for the subjects; otherwise, they would have been confronted with an additional decision problem, i.e. whether to purchase additional allowances or to initiate an abatement strategy. When the experiment is structured this way, the only decisions the subjects have to make concerns their activity in stages 1 and 2. This decision situation is repeated five times, i.e. the game consists of five periods, which are independent of each other. In addition, the game is played by groups consisting of six participants each. The treatments and subjects' characteristics are described in detail in the following section.

3.4.2 Treatments

The treatments differ with respect to the allocation process that takes place at the beginning of every period. This process regulates how to allocate an exogenously given total quantity of allowances among the six firms forming one group. Note that this total quantity of allowances represents the group's total emissions cap. These allowances are allocated free of charge (grandfathering) and/or via either a dynamic one-sided auction or a dynamic double auction, both with uniform pricing.[8] To determine which allocation configuration meets best the criteria described in Section 3.2, we test two hybrid scenarios and two single scenarios. The former utilize a combination of grandfathering (GF) plus either of the two auction types, i.e. the dynamic one-sided auction (A) or the dynamic double auction (DA). In the latter two scenarios, allocation is achieved by either grandfathering or the one-sided auction. All four scenarios are described in more detail below.

Hybrid Scenarios In Treatments $GF+A$ and $GF+DA$, the total quantity of allowances is allocated via a combination of grandfathering and auctioning. In both scenarios, subjects receive an initial stock of allowances by means of grandfathering. Then, in stage 1, each subject can acquire additional allowances by participating in an auction, either a one-sided auction ($GF + A$), where subjects can only buy allowances, or via a double auction ($GF + DA$), where subjects can both buy and sell allowances.

Single Scenarios In Treatment GF, the total quantity of allowances is completely allocated by means of grandfathering at the beginning of the period. In Treatment A, firms acquire allowances solely by a one-sided auction. Thus, because there is no auction in Treatment GF, the game starts directly with stage 2 (the market trading process), and in Treatment A, subjects are not granted an initial stock of allowances.

The requirement of comparability between the treatments is met by having the same total allocated quantity of allowances per period in all treatments. In the first period 1110

---
[8]In the following, the term auction refers to the dynamic uniform auction.

3.4. EXPERIMENTAL DESIGN

allowances are allocated, which decreases by 120 allowances in each successive period. Thus, in the fifth (last) period there are only 630 allowances left to allocate. Each subject emits 200 tons of CO_2 in every period. Thus, for each group of six participants, the total business-as-usual emissions sum up to 1200 tons of CO_2. Consequently, the total required abatement of allowances (i.e. the difference between total business-as-usual emissions and total allocated quantity) is the same per period in all treatments; in the first period the amount is 90 tons of CO_2, which increases to 570 tons of CO_2 in the last period. Thus, in every period all subjects face the same scarcity situation in all treatments (see Table 3.1).

Table 3.1: Basic characteristics for all treatments

Period	1	2	3	4	5
Total business-as-usual emissions [tons of CO_2]	1200	1200	1200	1200	1200
Total allocated quantity [CO_2 allowances]	1110	990	870	750	630
Total required abatement [tons of CO_2]	90	210	330	450	570

Note that we decided to change (reduce) the total allocated quantity from period to period in order to prevent subjects from deducing the scarcity conditions, which would be obvious if the total allocated quantity was the same in every period. Table 3.2 displays the breakdown of the total allowances allocated by grandfathering and/or auctioning according to the different treatments per period.

Table 3.2: Total allocated quantities via grandfathering and auctioning [in CO_2 allowances]

		Period				
	Treatment	1	2	3	4	5
Grandfathering	$GF + A$	960	840	720	600	480
	$GF + DA$	960	840	720	600	480
	GF	1110	990	870	750	630
	A	0	0	0	0	0
Auctioning	$GF + A$	150	150	150	150	150
	$GF + DA$	150	150	150	150	150
	GF	0	0	0	0	0
	A	1110	990	870	750	630

For the first period of the two hybrid treatments, i.e. $GF + A$ and $GF + DA$, 960 of the 1110 total allowances are grandfathered and the remaining 150 represent the exogenous auction supply. While the auction supply stays constant (150 allowances) over the course of the experiment, the grandfathered amount decreases by 120 in each successive period. The double auction (stage 1 of Treatment $GF + DA$) and trading process (stage 2 in all treatments) are both implemented as a dynamic uniform double auction where participants can act as buyers or sellers with the activity rule of a Japanese auction. That is, the auctioneer continuously raises the price until demand meets supply. At every price, participants have to signal their willingness to stay in the auction and to pay (receive) the current price for their demanded (offered) quantity.[9] The one-sided auction (stage 1 of Treatment $GF + A$) is implemented analogously but with a demand side, only. For the detailed implementation of the mechanisms we refer the reader to Appendix B.1.

3.4.3 Subjects' Characteristics

Table 3.3 summarizes the subjects' characteristics per period in all treatments. The individual characteristics for the two hybrid scenarios, Treatments $GF + A$ and $GF + DA$, do not differ.

As stated already, in every period all six subjects of a group face a given emissions commitment (business-as-usual emissions) of 200 tons of CO_2 (for a total of 1200 tons per group) that has to be covered by CO_2 allowances. Besides, except for Treatment A, the individual initial stock of allowances per period is determined by splitting the total grandfathered quantities equally among the six subjects of a group. For example, in the first period of Treatments $GF+A$ and $GF+DA$, the 960 grandfathered allowances yield an initial stock of 160 allowances per subject, which reduces to 80 allowances in the last period. Each subject also has an initial monetary endowment that increases over the course of the experiment, one abatement measure with a maximum abatement volume of 200 tons of CO_2, and con-

---

[9]This definition differs from Friedman (1991). He defines the double auction market as an institution, in which participants can as an institution in which make and accept public offers to buy (bids) and to sell (asks).

3.4. EXPERIMENTAL DESIGN

Table 3.3: Subjects' characteristics in the treatments

The table depicts subjects' characteristics per period in the treatments. Single entries denote that the respective characteristic stays constant in the course of the experiment, whereas a five dimensional entry represents the characteristic values for the first to the fifth period.

Treatment	Business-as-usual emissions per period [tons of CO_2]	Initial stock of CO_2 allowances from first to last period	Initial money endowment from first to last period [ExCU]	Abatement volume per period [tons of CO_2]
$GF + A$, $GF+DA$	200	160, 140, 120, 100, 80	800, 1200, 1600, 2000, 2400	200
GF	200	185, 165, 145, 125, 105	300, 700, 1100, 1500, 1900	200
A	200	0	800, 1200, 1600, 2000, 2400	200

stant MAC. In every period, the commonly known MAC distribution of $\{3, 6, 9, 12, 15, 18\}$ [ExCU per ton of CO_2] is assigned to the six subjects of a group such that in every period every number is only assigned once. Additionally, the MAC are reassigned from period to period such that each subject has different MAC in every period.[10]

With respect to the subjects' information structure, we distinguish between private and common information to stay as close as possible to the information structure of firms that participate in the EU ETS. The subjects' characteristics, i.e. their initial stock of allowances, monetary endowment, and MAC, are private information and change in every period.[11] All subjects know the size of their group and the number of periods, the total emissions commitment (total business-as-usual emissions) in every period, the total initial stock of allowances, the exogenous auction supply (if part of the allocation), and the distribution of the individual MAC. Individual and common characteristics are announced

---

[10]The individual monetary endowments are calculated such that they were sufficient for firms to acquire the number of allowances necessary to cover their emissions at the beginning of every period, i.e. only by abatement at the maximal possible MAC of 18 ExCU and without having to take part in the trading or auction processes. The distribution of individual MAC was designed to give all subjects approximately the same total abatement costs for the five periods so that they would receive approximately the same total profit in the theoretical overall cost minimum. Hence, each subject ends up in profitable situations with relatively cheap as well as with relatively expensive MAC, which is less profitable.

[11]Because we have modified the distribution of the MAC and the total allocated quantity, we cannot control for learning effects over the course of the experiment.

at the beginning of every period.[12]

The subjects' objective is to maximize their profit in every period. A subject's profit per period is given by his individual monetary endowment minus (plus) the value of the allowances purchased (sold) in the trading and/or auction process minus abatement costs. Note that surplus allowances become worthless at the end of every period. A subject's total profit in the game is determined by summing up his or her profits from every period.

3.4.4 Organization of the Experiment

We conducted the experiment at the University of Karlsruhe, Germany, where 156 students from various disciplines were randomly selected as participants. 18 subjects took part in each experimental session. Thus, for each treatment, we ran two sessions with three groups each (see Table 3.4). The experiment was computerized. Subjects received common written instructions, which were also read aloud by an instructor. Before the experiment started, each subject had to answer several questions about the instructions at his or her computer terminal. At the end of a session, subjects were paid in cash according to their total profits. For the instructions we refer to the Appendix B.3.

Table 3.4: Organization of the experiment

Treatment	Number of groups	Number of firms in each group	Initial allocation process of total quantity of allowances
$GF + A$	6	6	Grandfathering followed by a one-sided uniform auction
$GF + DA$	6	6	Grandfathering followed by a double uniform auction
GF	6	6	Exclusive grandfathering
A	6	6	Exclusive one-sided uniform auction

---

[12]This information structure basically enables participants to calculate bidding and abatement behavior in the cost minimum, i.e. according to the theoretical reference point; see Section 3.5.

3.5 Theoretical Considerations

In Appendix B.2 we provide the theoretical framework for the derivation of theoretical reference points for our experimental game. For this purpose, we develop a general two-stage model, which also includes our experimental design. Here, we distinguish between two behavior hypotheses: first, the firms play sophisticated equilibrium strategies, i.e. they take the whole process of grandfathering, auctioning, and trading into account and are able to correctly estimate the true market scarcity price; second, all firms behave myopically in the sense that they align their bidding strategies with their own MAC. Within this theoretical framework, we test our considered allocation rules with respect to their compliance with Criteria C1 to C3. Setting the focus on auctioning, we additionally distinguish between national auctions with a relatively small number of participants and one large (e.g. EU-wide) auction with many participants, in which a single firm's impact on the auction price can be disregarded. In the following we present the results of our theoretical analysis only. For a detailed elaboration we refer to Appendix B.2. The first result refers to the characteristics of equilibria strategies, whereas the second one refers to myopic bidding behavior only. For the equilibrium of the two-stage game, we can state the following result.

Proposition 1 Neither national (one-sided or double) auctions nor large (one-sided or double) auctions generate equilibria capable of fulfilling Criteria C1, C2, and C3 at the same time. However, in a large double auction in which participants engage in myopic bidding, an additional equilibrium is generated that simultaneously fulfills C1, C2, and C3. A large one-sided auction is also capable of fulfilling all three criteria at once, but only if firms do not possess any allowances at the time of the auction (i.e. no grandfathering or banking).

In other words, whereas national auctions of any sort fail to meet our criteria, certain kinds of large auctions are capable of fulfilling all three criteria. These include large double auctions in which myopic bidding strategies occur as well as large one-sided auctions in which firms do not possess any allowances at the time of the auction. However, none of the allocation rules under consideration is able to fulfill all criteria if firms play sophisticated equilibrium strategies, which take the entire form of the game (an auction followed by trading) into consideration. Since the derivation of these equilibrium strategies is a complex task for the firms, we additionally simplify the game by assuming myopic behavior.[13] That is, we now ask how things might turn out if we were to restrict firms to myopic bidding strategies, i.e. to straightforward bidding according to their individual MAC as postulated by Criterion C3. Our theoretical considerations in Appendix B.2 lead us to the following conclusion.

Proposition 2 If firms bid myopically, the one-sided as well as the double auction are expected to meet Criteria C1, C2, and C3 at the same time so long as the firms are not in possession of allowances. However, if firms possess allowances at the time of the auction (e.g. via grandfathering or banking), C1 to C3 are only fulfilled if the double auction format is applied. Both results are independent of the size of the auction.

Table 3.5 summarizes the compliance of the allocation rules applied in the four treatments with Criteria C1 to C3 in case of equilibrium and myopic bidding behavior. The presence of a dash instead of a check mark indicates that the respective criterion cannot be met unconditionally by the allocation rule. Note that as the criteria effectively concern the auction process only, no inferences can be made for Treatment GF.

With respect to the stability criterion for the auction process, the allocation mechanism $GF + A$ obviously has to be considered unstable if subjects bid myopically. In this case the auction price exaggerates the true market scarcity price, which is also assumed to be the price on the secondary market. If subjects realize that they have paid too much in the auction, it is assumed that they will adjust their bids in the next auction. In contrast, for the mechanisms $GF + DA$ and A, subjects notice that the price in the auction is equal to the one in the trading process and thus there is no incentive to change bidding strategies. Hence, both allocation rules are considered stable.

---

[13]Myopic bidding behavior is also supported by experimental studies, e.g., by Ehrhart et al. (2006).

3.5. THEORETICAL CONSIDERATIONS

Table 3.5: Evaluation of initial allocation rules

The table displays the evaluation of initial allocation rules with respect to the criteria information efficiency (C1), allocation efficiency (C2), and incentive compatibility (C3) in the presence of equilibrium and myopic bidding behavior.

Treatment	equilibrium			myopic		
	C1	C2	C3	C1	C2	C3
$GF+A$	✓	-	-	-	✓	-
$GF+DA$	✓	✓*	✓*	✓	✓	✓
A	✓	✓*	✓*	✓	✓	✓

* This applies to large auctions with many participants only, in which a single firm's impact of on the auction price can be disregarded.

Table 3.6 summarizes the theoretical reference prices derived from Propositions 1 and 2. With respect to information efficiency (C1), we conclude for the hybrid allocations $GF+A$ and $GF+DA$ that the double auction has to be considered superior to a one-sided auction because it generates correct price signals even if firms bid myopically. A one-sided auction format is only expected to yield good price signals if firms do not possess any allowances at the time of the auction (Treatment A).

Table 3.6: Sequence of reference prices in the treatments

Period	1	2	3	4	5
Scarcity price p^*	3	6	6	9	9
Trading price in all treatments	3	6	6	9	9
"Myopic" auction price in Treatment $GF+A$	9	12	15	15	15
"Myopic" auction price in Treatments $GF+DA$ and A	3	6	6	9	9
Equilibrium auction price in all treatments	3	6	6	9	9

For the analysis of our experimental data, the values in Table 3.6 serve as theoretical reference points for the auction and the market prices in every period. The derivation of the reference prices is illustrated in detail for the first period in Appendix B.2.

3.6 Experimental Results

In the following analysis we apply different statistical tests to our experimental data in order to assess the compliance of the four initial allocation rules with our stipulated Criteria C1 to C3. In Section 3.6.1 we compare prices and volumes in the auction and trading processes as well as abatement costs to investigate information efficiency (C1), the level of trading costs, and allocation efficiency (C2).[14] To investigate stability, in Section 3.6.2 we examine the incentive compatibility (C3) of the allocation mechanisms by analyzing subjects' bidding behavior.

3.6.1 Prices, Volumes, and Allocation Efficiency

Prices

When we analyze experimental prices, the relationship with the scarcity price p^*, which is based upon our hypothetical reference points in Section 5, comes to the fore. In the following, we calculate average prices by taking the mean of the prices in the five periods of all six groups per treatment. Figure 3.1 presents the sequence of average observed auction and trading prices together with the trajectory of the scarcity price p^* (see Table 3.6). An evident deviation from p^* can only be recognized in Treatment $GF+A$, where the price of the one-sided uniform auction exaggerates p^*. The exaggeration, however, tends to decrease over time.[15] The trading prices of all treatments stay relatively close to the sequence of p^*.

To analyze the quality of the observed prices we measure their deviations from the true market scarcity price p^* and test if they are significantly different from zero. Table 3.7 summarizes the average observed auction and trading prices together with their deviations from p^*, whereas deviations are presented both absolutely and relatively. The average

---

[14] Note that in the experiment we did not explicitly model transfer costs. Thus, we can only take the traded volume as an indicator for the degree of trading costs.

[15] A plausible explanation might be that subjects observed that they paid more in the auction than in the trading phase and thus bid more defensively in the next auction round. However, as mentioned in Section 3.4.2, our experimental design does not allow us to control for learning effects.

3.6. EXPERIMENTAL RESULTS

Figure 3.1: Average observed prices for each treatment

The figure displays average observed auction prices p_A, trading prices p_T, and scarcity prices p^* for each treatment.

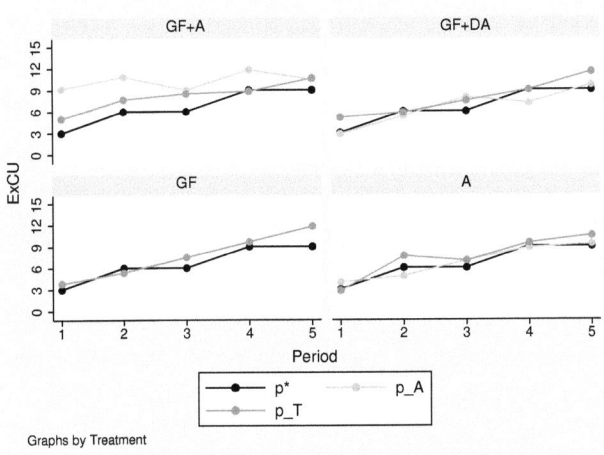

Table 3.7: Average observed prices and deviations from the cost minimum

The table depicts average observed auction and trading prices and their deviations from p^* in ExCU and percent (figures in brackets) for each treatment.

Treatment	Average Price [ExCU per allowance]			Deviation from p^* [ExCU]	
	p^*	p_A	p_T	Auction	Trading
$GF+A$	6.6	10.27	8.13	3.67 (56%)	1.53 (23%)
$GF+DA$	6.6	6.6	7.80	0.00 (0%)	1.20 (18%)
GF	6.6	-	7.63	-	1.03 (16%)
A	6.6	6.8	7.50	0.20 (3%)	0.90 (14%)

auction prices in Treatments $GF+DA$ and A are good price signals for p^*: the average auction price in $GF+DA$ exactly meets p^* and there is only a marginal deviation of 0.2

ExCU in Treatment A. Using the Wilcoxon rank-sum test, we find that neither deviation is significantly different from zero ($p > 0.45$ for $GF + DA$ and $p > 0.25$ for A).[16] In Treatment $GF + A$, however, we observe a statistically significant positive deviation of 3.67 ExCU (Wilcoxon rank-sum test: $p < 0.001$). When we compare the price deviations of the treatments, we obtain significantly higher auction prices in Treatment $GF + A$ than in Treatments $GF + DA$ and A (Kruskal-Wallis test: $p < 0.001$; Tukey test: $p < 0.05$ for both pairwise tests). Auction prices in Treatment $GF + DA$ do not differ significantly from those in Treatment A (Tukey test: $p > 0.05$).

With respect to average trading prices, we observe significant deviations from the scarcity price p^* in all treatments except for Treatment A (Wilcoxon rank-sum test: $p > 0.08$). However, compared to the auction price deviation in $GF + A$, these deviations are rather small (1.53 ExCU for $GF + A$, 1.20 ExCU for $GF + DA$, and 1.03 ExCU for GF,) which is also evident from the relative deviations. The trading prices of the four treatments do not show significant differences (Kruskal-Wallis test: $p > 0.08$).

Result 1 The auction design matters with respect to information efficiency. Only Treatments $GF + DA$ and A are able to create reliable price signals in the auction and thus meet Criterion C1.

Volumes

Table 3.8 displays the average observed trading volumes in the auction and trading processes together with the optimal volumes (figures in brackets), which are given by the lowest trading levels needed to achieve the optimum. Again, we average the five periods for all six groups. Because the auction volumes are exogenously given in Treatments $GF + A$ and A, we need only analyze the volumes in the trading process. We observe that significantly more allowances are traded in Treatments $GF + A$ and GF than in Treatment $GF + DA$ (Kruskal-Wallis test: $p < 0.001$; Tukey test: $p < 0.01$ for both pairwise tests). The ar-

---

[16] To receive statistically independent observations, we calculated the average of five periods of one group. As each treatment is played by six groups, we obtain a sample sizes of six units per treatment for the following statistical tests.

3.6. EXPERIMENTAL RESULTS

gument that the regular participation in the trading process induces higher transaction costs to the firms than occasional bidding in an auction Cramton and Kerr (2002); see also Criterion C2) thus constitutes an additional advantage of an allocation rule that stipulates a double auction such as Treatment $GF + DA$.

Table 3.8: Average observed and optimal volumes

The table depicts the average observed and optimal (figures in brackets) volumes in the auction and trading processes.

Treatment	Auction	Trading
$GF + A$	150 (150)	163 (154)
$GF + DA$	265 (154)	90 (0)
GF	-	185 (199)
A	870 (870)	145 (0)

Allocation efficiency

Criterion C2 requires allocation efficiency only for auctions. However, as we are studying ETS, we are also interested in the cost savings created by the possibility of trading emission allowances. For this purpose, we define indicator ψ as a general measure for allocation efficiency to control for Criterion C2 and to explore the extent to which the treatments realize cost savings compared to a system that does not permit the trading of allowances.

$$\psi = \frac{\text{AC(without ET)} - \text{AC(actual)}}{\text{AC(without ET)} - \text{AC(optimum)}} = \frac{\text{actual cost savings}}{\text{maximum cost savings}} \quad (3.1)$$

AC(without ET) denotes the minimum total abatement costs in the theoretical solution without ET, i.e. firms have to achieve their emissions commitment individually. AC(actual) is the sum of the actual ET abatement costs, and AC(optimum) represents the total abatement costs in the overall ET optimum. Thus, ψ is given by the ratio of actual and maximum cost savings and hence is an indicator for the achieved degree of efficiency in a group.[17] Therefore, the higher the value of ψ is, the higher the degree of cost-efficiency will be.[18] For our analysis we calculate ψ after the auction and again after trading in order to determine whether cost savings have increased. We assume that abatement takes place immediately after the auction and calculate the values hypothetically. The results are displayed in Table 3.9. Since in Treatment $GF + A$ only some of the allowances are available in the auctions, the figures in brackets additionally state the percentage of efficiently allocated allowances within the auction supply, i.e. the percentage of the auction supply that is bought by those subjects with the highest MAC.

Table 3.9: Degree of cost-efficiency

The table depicts the degree of cost-efficiency ψ in every period after the auction and after trading in all treatments.

Treatment	Measured after	Period					Average
		1	2	3	4	5	
$GF + A$	Auction	-0.13	0.05	0.13	0.31	0.16	0.10
		(-0.17)	(0.10)	(0.29)	(0.80)	(0.41)	(0.29)
	Trading	0.19	0.63	0.87	0.75	0.89	0.67
$GF + DA$	Auction	0.34	-0.10	0.23	0.24	0.69	0.28
	Trading	0.47	0.28	0.67	0.60	0.95	0.59
GF	Trading	-0.12	0.58	0.77	0.82	0.80	0.57
A	Auction	-2.39	-0.64	0.17	0.47	0.61	-0.35
	Trading	-0.91	0.38	0.73	0.88	0.82	0.38
Average	Auction	-0.73	-0.23	0.18	0.34	0.49	0.01
	Trading	-0.09	0.47	0.76	0.76	0.87	0.55

Taking the average of all six groups and all five periods, the degree of efficiency ψ after the auction (trading) is equal to 0.01 (0.55), while the values range from -2.39 to 0.69 (-0.91

---

[17]The cost-efficiency measure ψ is a well-established measure typically used in ET experiments. For a survey of experimental ET efficiency values, see Muller and Mestelman (1998), p. 230, Table 2.
[18]The maximum value ($= 1$) is reached in the overall optimum, where 100% of the potential cost savings are achieved. A negative value indicates that the group performs worse than in the theoretical solution without ET.

3.6. EXPERIMENTAL RESULTS 101

to 0.95).[19] In all treatments ψ increases from auction to trading and across the periods, and on average, slightly more than half of the potential cost savings are realized after the trading stage. In the last period after the trading stage all treatments achieve at least 80% of the potential cost savings. This suggest that subjects tend to become acquainted with the ETS during the experiment. Consequently, we believe that Period 5 ought to be considered the most representative period: at that point, subjects can be assumed to have become familiar with the ET game, and furthermore, the subjects have to exert the most activity in this period in order to reach their total emissions commitment. Thus, the following result is derived.

Result 2 Emissions trading leads to greater cost savings compared to the theoretical solution without emissions trading. In the last period, a large part of these savings is already achieved by the auction in Treatments $GF + DA$ (69%) and A (61%), as required by Criterion C2.

3.6.2 Bidding Strategies

We now analyze subjects' bidding behavior with respect to the reservation price for their buying and selling strategies in the auction and trading stages. According to our hypothesis in Section 3.5, subjects who play the equilibrium strategy are expected to be geared towards the scarcity price p^*, while myopically behaving subjects bid according to their individual MAC. First, we study the price when buyers decide to abandon the buyer position in the auction, i.e. when they drop out. Second, we investigate if sellers offer allowances at a market price that is above their individual MAC or above p^*.

---

[19]Note that a negative degree of efficiency after the auction (especially in Treatment A) does not necessarily imply an inefficient allocation mechanism. Due to the construction of the measure 3.1, subjects' excess allowances do not figure into the calculation. In other words, for those subjects possessing more allowances than necessary for their CO_2 commitment after the auction, we are not able to distinguish between subjects who intend to speculate by selling these excess allowances later in the secondary market and those who simply misunderstand the ET game and thus cause inefficiencies.

Buyer Dropouts

We call the price at which subjects decide to stop submitting a purchase bid the dropout price. Since a subject in the seller position is locked into this role and cannot abandon it, this analysis focuses on buyers only. Dropouts are defined as subjects who either quit the process altogether or switch from the buyer to the seller position.

When utilizing the myopic strategy, a subject drops out of the auction when the current auction price reaches his or her MAC or MAC+1. At these prices, he or she is indifferent between buying allowances in the auction and abating emissions. Dropping out at price p^* or p^*+1 characterizes the equilibrium bidding strategy, because the subject expects to buy or sell allowances at p^* in the after market. Figure 3.2 displays histograms of the absolute deviations of the dropout price p from the two different reservation prices for the auction in Treatments $GF + A$, $GF + DA$ and A. A glance reveals that there is always a higher peak at MAC than at p^*. Moreover, the variance is higher for the equilibrium reference point p^*.

Table 3.10: Percentage of dropouts in the auction that are consistent with individual MAC and p^*

Treatment	Auction [%] MAC	p^*	Trading [%] MAC
$GF + A$	43	27	52
$GF + DA$	35	15	54
GF	-	-	63
A	47	38	47
Average	42	27	54

The second and third columns in Table 3.10 show the percentage of dropouts in the auctions that are consistent with MAC and p^*. In Treatment $GF + A$, 43% of the dropout decisions in the auction are in line with the individual MAC (at MAC or MAC+1), whereas 27% of the dropouts are based on p^* (at p^* or p^*+1). As the table indicates, the other treatments show similar results. By statistically comparing the percentages we find that in Treatments

3.6. EXPERIMENTAL RESULTS

Figure 3.2: Histograms of dropout prices in the auction

The figure displays histograms of absolute deviations of dropout price p from individual MAC, i.e. MAC-p (upper graphs a–c) and from p^*, i.e. p^*-p (lower graphs d–f) in the auctions.

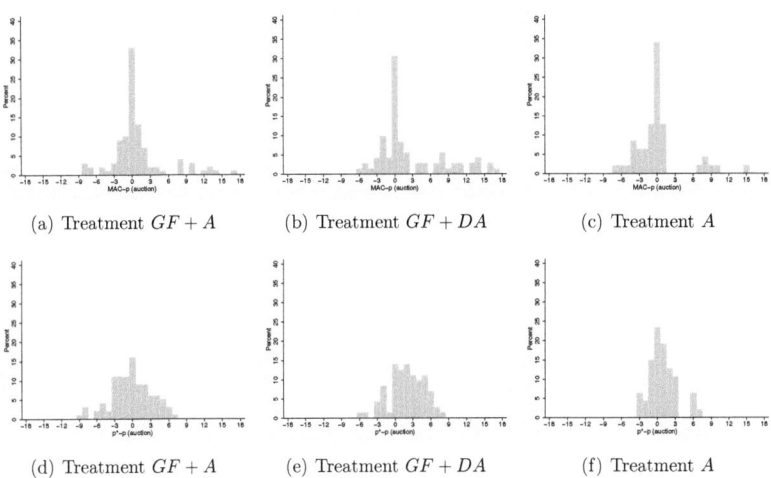

(a) Treatment $GF + A$ (b) Treatment $GF + DA$ (c) Treatment A

(d) Treatment $GF + A$ (e) Treatment $GF + DA$ (f) Treatment A

$GF + A$ and $GF + DA$ significantly more decisions are in line with the MAC than with p^* (Wilcoxon rank-sum test, $p < 0.05$).

Additionally, the last column of the table shows the percentage of dropout decisions according to MAC in the trading process, which constitutes an equilibrium strategy if a single buyer's impact on the market price is negligible. It can be seen that on average, more than half of all dropouts in the trading process are consistent with subjects' MAC.

Result 3 In the auction and trading processes the majority of subjects initially submits purchase bids according to individual MAC rather than to the scarcity price p^*.

Sellers in the Double Auction and Trading

We analyze the behavior of subjects who intend to sell allowances in the double auction of Treatment $GF + DA$. We begin by considering the price at which a subject submits a selling offer for the first time, which is then valid until the end of the process. This price reflects the subject's willingness to accept.

As in the previous analysis, we consider a subject's strategy to be myopic if the offer is submitted for the first time at a price level of MAC or MAC+1. Analogously, the equilibrium bidding strategy is characterized by first selling offers set at p^* or $p^* + 1$. Figure 3.3 shows histograms of the absolute deviations of the price of the first selling offer p from the two different reservation prices for the double auction in Treatment $GF + DA$. Again, the graphs clearly suggest myopic selling behavior.

Figure 3.3: Histograms of first selling offers in the auction

The figure displays histograms of absolute deviations from the price of the first selling offer p to individual MAC, i.e. MAC-p (left graph) and from p^*, i.e. p^*-p (right graph) in the auctions.

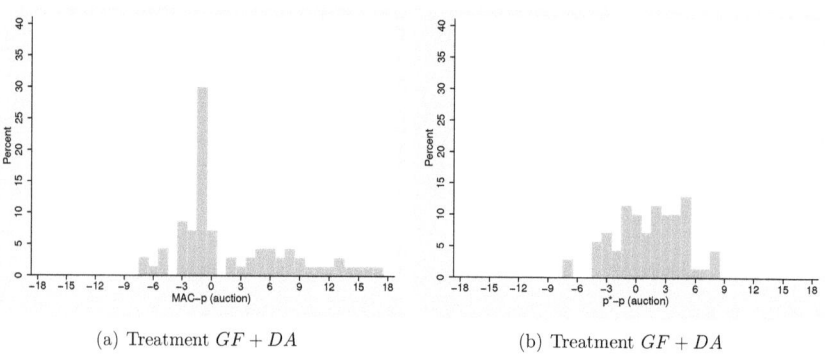

(a) Treatment $GF + DA$ (b) Treatment $GF + DA$

Table 3.11 gives the percentage of the first-time offers that are consistent with the two reference prices for the double auction in Treatment $GF + DA$ and for the trading phase (stage 2) in all treatments. With respect to the double auction, the statistical analysis

3.6. EXPERIMENTAL RESULTS

confirms the conclusion drawn from the graphs: the percentage of MAC-based offers (37%) is statistically higher than the percentage of p^*-based offers (21%) (Wilcoxon rank-sum test: $p < 0.05$).

Table 3.11: Distribution of submitted offer prices

The table displays the distribution of submitted offer prices p that are consistent with MAC and p^*.

Treatment	Auction [%] MAC	p^*	Trading [%] MAC
$GF + A$	-	-	42
$GF + DA$	37	21	38
GF	-	-	44
A	-	-	15
Average	37	21	35

Looking at the trading process, we observe that on average, 35% of the subjects start to offer at their own MAC. However, we find that many selling bids are submitted at a price level below the individual MAC. In Treatment A, for example, such bids constitute more than half of the offers (52%). To shed some light on this phenomenon, we additionally differentiate between sellers who still require allowances to fulfill their emissions commitment when entering the trading stage and those who don't. This gives us a clearer picture: most of the sellers who offer at a price below their MAC have already fulfilled their emissions commitment in the auction and thus possess more allowances than they actually need (81%). Since these allowances become worthless after trading, these subjects have a strong incentive to sell them at any price, even at a price that is lower than their MAC. Most of the subjects who still require allowances after the auction (only) sell at a price above or equal to their MAC (90%).

Result 4 In the double auction the majority of subjects submits selling bids according to their individual MAC. In the trading process, subjects tend to try to minimize their losses by offering surplus allowances at any price, even at prices that are lower than their MAC.

Results 3 and 4 strongly support the hypothesis of myopic bidding behavior in the auction process. When these two results are combined with Result 1, which shows that in Treatments $GF + DA$ and A the auction prices do not differ significantly from the scarcity prices p^*, we conclude that subjects do not have an incentive to deviate from their myopic bidding strategy over the course of the experiment, i.e. the two allocation mechanisms are considered to be stable. Thus, Criteria C3, i.e. incentive compatibility, is indirectly supported. However, this does not hold for Treatment $GF + A$, where subjects observe auction prices that significantly exaggerate p^*.

3.7 Conclusion

In our paper we emphasized the importance of a properly chosen initial allocation rule for the successful implementation of the EU ETS and other future CO_2 trading schemes. We believe the implementation will be successful if the allocation rule meets three criteria: information efficiency, allocation efficiency, and incentive compatibility. Information efficiency (C1), or the creation of early and reliable signals, helps firms to make correct abatement decisions. Allocation efficiency (C2) leads to the funneling of allowances to the firms that need them most, i.e. to the firms with the highest abatement costs. Finally, incentive compatibility (C3) enables firms to simply base their bidding strategies on their own abatement costs. The objective of this paper was to find an initial allocation rule for Phase 3 of the EU ETS that embodies these three criteria. We obtained our results by comparing several relevant allocation rules in a theoretical and experimental framework.

The most outstanding result of our theoretical and experimental analysis is that the allocation rule favored by most EU countries for Phase 3 and by other international ET schemes, i.e. grandfathering in combination with a one-sided uniform auction, does not in fact fulfill the proposed criteria. This rule does not promote the straightforward bidding postulated by C3. Instead, it encourages behavior that results in auction prices that exceed the scarcity price and thus does not create reliable signals. Reliable signals can only be generated if firms take the whole process into consideration. However, participation

3.7. CONCLUSION

then becomes even more complex, and additionally, the postulate of cost-efficiency would likely be violated. Besides, if there is no complete positive correlation between individual MAC and signals, the auction allocation is not expected to be efficient. Since exclusive gratis allocation (grandfathering) also yields inefficiencies, we proposed the following two allocation rules: If grandfathering is to be part of the allocation process – as planned for the industry sector in Phase 3 – the regulator might be on the "safe side" by applying a uniform double auction. If exclusive auctioning is possible – as might be the case for the electricity sector after 2012 – then a uniform price one-sided auction is feasible. If we assume firms to behave myopically, as observed in the laboratory experiment, both rules fulfill Criteria C1 and C2. In addition, since the stability criterion holds for both rules, we consider Criteria C3 to be met also. Thus, we argue that among a pool of several initial allocation mechanisms, these two scored well and seem to be attractive candidates for Phase 3.

Additionally, with respect to double auctions, our result may also be used as an argument for the German implementation of allowed auctioning for the years 2008 and 2009 of Phase 2. Here, the governmental auction supply is offered in equal portions every day on the trading platforms of the secondary market at the market price. Obviously, when it comes to practice, the advantages and disadvantages of both institutions need to be evaluated and tested experimentally in more detail.

Appendix

A Appendix to Chapter 2

A.1 Components of Estimated Half Spreads

Table A.12: Components of estimated half spreads

The table depicts the estimated transitory component $\hat{\phi}$ and permanent component $\hat{\theta}$ in Euro for ECX and Nord Pool obtained by GMM estimation of Equation (2.3) under the given moment conditions. Estimation periods are as indicated with e.g. Q3 2006 denoting July to September 2006. For Nord Pool and ECX, the last estimations for the Dec05 contract are from September to December of the respective year. For ECX, estimates for the third quarter 2007 (Q3 2007) are from June to September 2007. Results with less than 100 observations are not depicted. *,**, and *** denote statistical significance at the 10, 5, and 1 percent levels, respectively.

Contract		ECX				Nord Pool			
		$\hat{\phi}$	t-stat.	$\hat{\theta}$	t-stat.	$\hat{\phi}$	t-stat.	$\hat{\theta}$	t-stat.
Dec05	2005	-0.0024	-0.60	0.0648***	13.95	-0.0110	-1.10	0.0860***	7.36
	Q2	-0.0153**	-2.14	0.0767***	9.08	-0.0113	-1.01	0.0708***	5.59
	Q3	-0.0010	-0.17	0.0780***	11.23	-0.0104	-0.68	0.1037***	5.78
	Q4	-0.0005	-0.11	0.0423***	8.97	-0.0078	-0.72	0.0735***	5.17
Dec06	2006	0.0016	0.94	0.0496***	20.20	-0.0008	-0.07	0.0898***	7.89
	Q1	0.0036	1.62	0.0412***	15.75	-0.0135	-1.32	0.0945***	7.19
	Q2	-0.0037	-0.78	0.0834***	13.23	0.0095	0.35	0.1157***	4.79
	Q3	0.0019	0.93	0.0342***	13.61	0.0103	1.60	0.0484***	5.20
	Q4	0.0059***	2.92	0.0243***	9.14	-0.0213**	-2.22	0.0560***	4.35
Dec07	2007	0.0018*	1.80	0.0159***	13.02	0.0015	0.20	0.0231***	3.63
	Q1	0.0004	0.28	0.0210***	12.96	-0.0010	-0.10	0.0332***	3.83
	Q2	0.0043***	3.03	0.0075***	5.59				
	Q3	0.0031**	2.53	0.0028***	2.75				
Dec08	2007	0.0025***	5.85	0.0241***	46.30	-0.0064**	-2.20	0.0628***	17.04
	Q1	0.0054***	3.76	0.0350***	21.62	-0.0075	-0.77	0.0851***	7.87
	Q2	0.0009	0.82	0.0312***	25.39	-0.0123**	-2.16	0.0819***	11.28
	Q3	0.0027***	4.76	0.0188***	28.36	0.0052	0.87	0.0556***	7.61
	Q4	0.0020***	3.51	0.0186***	28.36	-0.0091**	-2.39	0.0407***	10.28

B Appendix to Chapter 3

B.1 Design of Dynamic Uniform Double Auction

In the experiment the trading and auction process are implemented as a dynamic uniform double auction. Players simultaneously submit their supply or demand bids of units in the form of a quantity bid at an initial price $p = 1$ ExCU. If the total demand bids exceed the total supply bids, the current price is increased by 1 ExCU and a new bidding round starts. The bidding continues until total demand is less or equal than total supply. The units are then allocated at the price of the last or the round before last. This depends on whether total demand in the last round was equal or smaller than total supply. Those buyers are rationed who reduced their quantity in the last round. The activity rule says that each buyer cannot increase and each seller cannot decrease his quantity as the price rises. Hereby we already equip subjects with monotone bidding strategies which help to bid rationally and prevent from absurd bidding behavior. During the trading process a buyer can always switch to a seller position or drop out completely from the trading process whereas this is not possible for the seller position. Once a selling bit is submitted it is valid until the auction and trading process is over.

B.2 Two-Stage Model

Let us consider a representative country with n committed risk-neutral firms.[20] Firm $i \in \{1, \ldots, n\}$ produces business-as-usual emissions (i.e. without abatement activity) of $e_i > 0$ and possesses an initial stock of allowances $s_i \geq 0$. Let $e = \sum_i e_i$ denote the total emission volume and $s = \sum_i s_i$ the total stock of allowances. Note that the case of grandfathering is covered by $s > 0$. Thus, all statements that refer to a constellation with grandfathering (i.e. Treatments $GF + A$ and $GF + DA$) also apply to situations where firms enter the auction with allowances in their possession, e.g. by banking.

---

[20]Since we consider an international competitive trading environment, we restrict the analysis to one representative country.

Firms are able to abate emissions and to trade allowances in an auction and on the market. If firm i abates volume a_i, it suffers costs according to its individual abatement costs functions $c_i(a_i)$ for which the common properties apply: $c_i(0) = 0$ and $c_i(a_i) > 0$ for $a_i > 0$ as well as $c'_i(a_i) = \partial c_i(a_i)/\partial a_i \geq 0$ and $c''_i(a_i) = \partial^2 c_i(a_i)/\partial a_i^2 \geq 0$ for all $a_i \geq 0$.

We use a two-stage model. In stage 1, allowances are traded in an auction with an exogenous supply of q allowances, where q_i denotes the number of allowances traded by firm i in the auction, $\sum_i q_i = q$.[21] Several kinds of auctions are possible: e.g. national auctions in each member state or even one EU-wide auction. In stage 2, firms trade allowances on the international market and decide on their abatement volume. Thus, the market price is uncertain at stage 1. This model is in line with the experimental setup. Solving the game by backward induction yields the following optimization problem.

In stage 2, ET takes place in an international competitive market where the firms are assumed to act as price takers and trade the allowances at a price p^*. Let $z_i := e_i - s_i - q_i$ denote the proportion of firm i's emissions volume that is not covered by allowances before trading, and d_i the quantity of allowances firm i trades on the market, where $d_i > 0$ indicates buying and $d_i < 0$ selling allowances, respectively. Hence, firm i's abatement requirement is given by $a_i = z_i - d_i$. Minimizing i's costs $c_i(a_i) + (z_i - a_i)\,p^*$ yields the first order condition

$$c'_i(a_i) = p^*, \qquad (B.1)$$

which represents the well-known condition that all firms' MAC are equal to the equilibrium market price p^* in a competitive market (e.g., Dales (1968); Montgomery (1972)). Price p^* expresses the international scarcity of allowances, and we additionally assume that p^* also reflects the scarcity conditions on the national market of our representative country. Firm i's optimal abatement volume at p^*, derived from condition (B.1), is denoted by $a_i(p^*)$.

In stage 1, q allowances are offered in a national or international auction. In the (one-sided or double) auction, firm i carries out its bidding function $b_i : \mathbb{R}^+ \to \mathbb{R}^+$, which indicates

---

[21] We assume that the whole exogenous auction supply is sold in the auction.

the quantity $b_i(p)$ firm i wants to buy (sell) at a certain auction price p. We impose the restrictions $b_i(\cdot) \in [0, \bar{q}]$ in a one-sided auction and $b_i(\cdot) \in [-\bar{q}, \bar{q}]$ in a double auction, $\bar{q} > 0$.

Equilibrium Bidding Behavior

In the following, we provide a general game-theoretic approach for the analysis of our extensive form game, in which an initial auction is followed by a perfect secondary market. This approach allows for the evaluation of the considered initial allocation rules with respect to Criteria C1 to C3.

By deriving its equilibrium strategy, a firm takes the whole process of grandfathering, auctioning, and trading into account. In the (single or double) auction, which is followed by trading, firm i's bidding strategy b_i together with the combination of other firms' bidding strategies b_{-i} determines the quantity of allowances traded by firm i and its payment in the auction, denoted by $q(b_i, b_{-i})$ and $r(b_i, b_{-i})$, respectively. Since the market price p^* is unknown at the time of the auction, firm i's expected total costs $TC(\cdot)$ are given by

$$E[TC(b_i, b_{-i})] = E[r(b_i, b_{-i}) + c_i(a_i(p^*)) + (e_i - s_i - a_i(p^*) - q(b_i, b_{-i})) \cdot p^*]. \quad (B.2)$$

Sorting Equation (B.2) with respect to b_i leads to

$$E[TC(b_i, b_{-i})] = E[r(b_i, b_{-i}) - q(b_i, b_{-i}) \cdot p^*] + E[K] \quad (B.3)$$

with $K := c_i(a_i(p^*)) + (e_i - s_i - a_i(p^*)) \cdot p^*$.

Generally, a (Bayesian) equilibrium of the auction is characterized by a combination of bidding functions $(b_1^*, \ldots, b_n^*)$, which simultaneously minimize each firm's expected costs. Formula (B.3) reveals two important properties. First, the solution of the cost minimization problem and thus a firm's equilibrium bidding strategy does not (directly) depend on the firm's abatement costs. Second, there is a direct dependency on the future market price p^*, which is the same for all firms and unknown at the time of the auction. Thus, the allowances in the auction take on the properties of a common value good, which has to be taken into consideration by the firms.[22] Calculating the Bayesian equilibrium presupposes assumptions about bidders' information structure, i.e. the distribution of individual information (signals) about the scarcity price p^* (see, for example, Milgrom (1981); Pesendorfer and Swinkels (1997)). However, it will become clear in the following that our general model (B.2) already suffices to evaluate the allocation mechanisms with respect to the three criteria. We find one plausible property of bidding strategies that applies to our approach and helps to evaluate the allocation mechanism: if a symmetric Bayesian equilibrium consisting of monotonic bid functions (with respect to the individual signals) exists, then an efficient auction outcome (C2) presupposes that signals and MAC are completely positively correlated, e.g. a firm's signal is equal to its MAC. However, playing a Bayesian equilibrium would present the firms with a big challenge as it requires outside knowledge, which is not incentive compatible (C3).

We assume that in addition to knowing their own MAC, the firms have access to other valuable information that at least allows them to determine if their abatement costs are e.g. above or below average. In our stylized experimental framework, each firm is informed about the distribution of MAC and the scarcity condition, i.e. the difference between the total emission volume without abatement and the total allocated quantity of allowances. Hence, the bidder is in principle able to compute the scarcity price p^*, which serves as the best signal for all firms, independent of their own MAC. In this game there exists a symmetric Bayesian equilibrium where all bidders apply the same bidding function $b^*(\cdot)$, which stipulates bidding $b^*(p) = \bar{q}$ for all $p \leq p^*$ and, in case of a double auction, $b^*(p) = -\bar{q}$ for all $p > p^*$. On the one hand, these equilibrium strategies are simple and since the auction price is expected to meet the scarcity price, the equilibrium is expected to generate a reliable price signal. On the other hand, although carrying out the equilibrium strategies seems to be an easy task, the firms in fact encounter difficulty in computing their guiding

---

[22] As a matter of course, emission allowances have the characteristics of private value goods. However, taking into account the resale opportunity on the market, the value of the allowance is determined by the market price, which is the same for all firms.

value p^* because since a lot of information has to be gathered and processed (distribution of MAC, total market scarcity). Since this process and the strategies themselves are assumed to be (largely) independent of bidders' individual MAC, the incentive compatibility criterion C3 is considered to be unmet. As a result, the auction mechanism is generally incapable of providing an efficient allocation (C2).

Our theoretical analysis is summarized by Proposition 1 in Section 3.5. Proposition 1 states that the equilibrium of a (one-sided or double) auction generally cannot fulfill Criteria C1, C2, and C3 at the same time. However, there is one exception (see also next paragraph): in the case of a large double or exclusive one-sided auction with many participants, in which the impact of a single firm on the auction price can be neglected, there exists an additional equilibrium in myopic bidding strategies (i.e. straightforward bidding according to individual MAC).

Myopic Bidding Behavior

A myopically bidding firm behaves in the auction in the same manner as on the market in stage 2, i.e. it restricts its bidding strategy to its individual MAC. In other words, it uses its MAC as a reserve bidding price. As a consequence, if all firms behave in this way, the auction already provides for an efficient outcome (C2). Hence, we have to test for information efficiency (C1). Additionally, we test for incentive compatibility (C3) in an indirect and weak form in the sense of the stability criterion by examining a firm's incentive to deviate from myopic bidding behavior.

We illustrate myopic bidding in our experimental framework, where each firm has one abatement measure with constant MAC. A myopic firm demands its whole shortage (i.e. the difference between its emission volume and its stock of allowances) as long as the price is below its MAC. If the price exceeds its MAC, the firm stops bidding in the one-sided auction or switches to offering its whole stock of allowances in the double auction, respectively.

Let us exemplarily consider the first period of our experimental game (Table B.13). Here, a total of 90 tons of CO_2 (the difference between the business-as-usual emissions of 1200 tons of CO_2 and the total allocation quantity of 1110 tons of CO_2) have to be abated. Since the maximum abatement volume of each measure equals 200 tons, the target is achieved cost-efficiently by the cheapest abatement measure (owned by subject 6). Thus, the lowest MAC determine the scarcity price, i.e. $p^* = 3$ ExCU per ton of CO_2.

What prices arise when firms bid myopically? In the two hybrid systems, each of the six firms is endowed with 160 allowances (i.e. 960 grandfathered allowances in total), and an additional 150 allowances are offered in the auction, which bring the total sum to 1110. In Treatment A this amount is exclusively sold via the one-sided auction. If subjects bid myopically by taking their individual MAC as reservation price, this yields the following outcomes:

- In Treatment $GF + A$, each firm i bids for $e_i - s_i = 40$ allowances as long as the price is below its MAC, which leads to an auction price of $p_A = 9$ ExCU per allowance. The auction supply of 150 allowances is then allocated to firms 1, 2, 3 (each receives $q_i = 40$ allowances), and 4 (which receives $q_4 = 30$). Firm 4 then determines the auction price $p_A > p^*$. Note that if the firms consider $p_A = 9$ as a correct market price signal, firms 4, 5, and 6 have an incentive to abate their missing quantities after the auction (i.e. $a_4 = 10$, $a_5 = 40$, $a_6 = 40$) at their individual MAC of 9, 6, and 3 ExCU per ton, respectively. This behavior prevents cost-efficiency, which is only achieved when firm 6 single-handedly abates the remaining 90 tons.

- In $GF + DA$, the auction supply of 150 allowances is allocated as before to firms 1, 2, 3, and 4. Additionally, firm 6, with the cheapest abatement technology, offers its whole stock of allowances when the price reaches its MAC of 3 ExCU and thus sells 50 allowances to firms 4 and 5 ($q_4 = 40$, $q_5 = 10$), which leads to $p_A = p^* = 3$.

- In Treatment A, each firm demands 200 allowances as long as the price is below its MAC, which leads to $p_A = 3$. The auction supply of 1110 tons is allocated to firms 1 through 5 ($q_i = 200$) and 6 ($q_6 = 110$). Firm 6 then determines the auction price.

B. APPENDIX TO CHAPTER 3   117

Table B.13: Example of myopic bidding behavior

The table depicts auction price p_A, initial stock of allowances s_i, purchasing auction volume q_i, and abatement volume a_i of myopically bidding firms in the first period of Treatments $GF + A$, $GF + DA$, and A

Firm		$GF + A$			$GF + DA$			A		
i	e_i	MAC_i	s_i	q_i	a_i	s_i	q_i	a_i	s_i	q_i	a_i
1	200	18	160	40	0	160	40	0	0	200	0
2	200	15	160	40	0	160	40	0	0	200	0
3	200	12	160	40	0	160	40	0	0	200	0
4	200	9	160	30	10	160	40	0	0	200	0
5	200	6	160	0	40	160	40	0	0	200	0
6	200	3	160	0	40	160	-50	90	0	110	90
Auction price p_A			9			3			3		

To summarize, myopic bidding provides for an efficient auction outcome (C2) for all auction formats independently of whether firms dispose of allowances or not at the time of the auction. However, the reliability of the price signal (C1) crucially depends on the latter condition. If the firms do not possess allowances at the time of the auction (i.e. all emission allowances are allocated via the auction), this requirement is met for a one-sided as well for a double auction. Otherwise (e.g. with grandfathering before the auction), this is only true for the double auction, whereas the signal of its one-sided counterpart is expected to exaggerate the scarcity price p^*.

It is also helpful for our analysis to consider a case in which firms are sufficiently small and a single firm's impact on the auction price can be disregarded. This would in fact occur if the European Commission were to decide on one EU-wide auction. If in this case all other firms bid myopically in the double auction, a single firm's set of best reply strategies also contains the myopic bidding strategy (as well as the equilibrium strategy described in Section B.2), independent of the firm's stock of allowances (Treatment $GF + DA$). In this case a myopically bidding firm has no incentive to unilaterally deviate from this strategy. Thus, myopic bidding constitutes an additional symmetric equilibrium, in which allowances are uniformly traded at price p^*. On this account, we generally consider the format of the double auction as (almost) incentive compatible (C3). These arguments also apply to the one-sided auction, but only if it is exclusively used for allocation (Treatment A). If firms possess allowances at the time of a one-sided auction (Treatment $GF+A$), myopic bidding leads to a higher price than p^*, and thus a firm with MAC $> p^*$ has an incentive to bid less than its MAC. Thus, the stability criterion has to be considered as violated and Criterion C3 is not fulfilled.

Recapitulating, these considerations allow us to draw the conclusion that under the assumption of myopic bidding, the one-sided as well the double auction are expected to meet Criteria C1, C2, and C3 at the same time so long as the firms are not in possession of allowances. However, if firms possess allowances at the time of the auction (e.g. via grandfathering), C1 to C3 are only fulfilled if the double auction format is applied. This result is presented as Proposition 2 in Section 3.5.

B.3 Instructions

In the following we provide the instructions for Treatment $GF+A$ only, as the three other treatments simply consist of elements of this treatment: trading (stage 2) is organized equally in all treatments as a dynamic uniform double auction (see Appendix B). Consequently, the instructions for the double auction in Treatment $GF+DA$ is analog to the one for the trading process. Obviously, at the end of the auction process subjects pay the auction price instead of the market clearing price.

In designing our experiment we decide against using ecological terms. We introduce a "neutral good", labeled by X, and replace the firms' CO_2 emissions commitment by a delivery commitment of a given quantity of X. The units of X can be traded among the participants. Displaying the initial allocation process, we either endow firms with X, which is analogous to the initial stock of allowance in case of grandfathering, and/or conduct an auction. In order to capture the emission abatement possibility, we allow each firm to produce units of X by itself.

Anleitung

Sie nehmen an einem Experiment teil, in dem Sie abhängend von Ihren Entscheidungen und den Entscheidungen der anderen Teilnehmer bares Geld verdienen können, das Ihnen am Ende des Experiments in Euro und anonym ausbezahlt wird. Die Recheneinheiten in diesem Experiment sind sogenannte Geldeinheiten (GE), wobei 250 GE einem Euro entsprechen. Jeder Teilnehmer trifft seine Entscheidungen unabhängig von den anderen Teilnehmern an seinem Computerterminal. Kommunikation zwischen den Teilnehmern ist nicht erlaubt.

Ausgangssituation

Sie bilden zusammen mit fünf weiteren Teilnehmern eine Sechsergruppe. Die Zusammensetzung der Gruppe bleibt für den ganzen Verlauf des Experiments dieselbe und es gibt keine Interaktion zwischen den Gruppen. Das Experiment läuft über 5 Perioden. In jeder Periode stehen Sie und die anderen 5 Teilnehmer der gleichen Entscheidungssituation gegenüber. In jeder der 5 Perioden repräsentieren Sie sowie auch jeder andere Teilnehmer in Ihrer Gruppe ein Unternehmen. Am Ende einer jeden Periode haben Sie die Verpflichtung, insgesamt 200 Mengeneinheiten (ME) eines Gutes X an Ihre Kunden auszuliefern. Zu Beginn einer jeden Periode verfügen Sie bereits über einen Anfangsbestand des Gutes X. Um die Lieferverpflichtung zu erfüllen, können Sie zusätzliche ME entweder kaufen oder selbst produzieren. Sie können jedoch auch ME verkaufen:

- Zunächst haben Sie in einer Auktion die Möglichkeit, zusätzliche ME des Gutes X zu kaufen.

- Im Anschluss an die Auktion findet ein Handel statt, in dem alle 6 Teilnehmer untereinander handeln können. D.h. Sie können ME des Gutes X von anderen Teilnehmern kaufen bzw. an andere Teilnehmer verkaufen.

- Nach dem Handel haben Sie die Möglichkeit, das Gut X in Eigenproduktion selbst zu produzieren. Dies geschieht zu Ihrem individuellen Produktionspreis je ME. Dieser wird Ihnen zu Beginn einer jeden Periode auf Ihrem Bildschirm bekannt gegeben. Ihre Firma verfügt in jeder Periode über eine Produktionskapazität von 200 ME des Gutes X

Bitte beachten Sie, sollte ihr gesamter Produktionsbestand am Ende einer Periode die auszuliefernde Menge von 200 ME übersteigen, so hat dies keine Konsequenzen für Sie. Jedoch werden die überschüssigen ME für Sie wertlos.
In jeder der 5 Perioden verfügen Sie über eine finanzielle Grundausstattung. Diese ermöglicht Ihnen, in der Auktion und im Handel ME des Gutes X zu kaufen bzw. in Eigenproduktion selbst zu produzieren.
Im Folgenden wird der eben zusammengefasste Ablauf des Experiments detailliert beschrieben.

Anfangsbestand und finanzielle Grundausstattung

Zu Beginn einer jeden Periode verfügen Sie über einen Anfangsbestand des Gutes X (in ME) und eine finanzielle Grundausstattung (in GE). Der folgenden Tabelle können Sie für jede der 5 Perioden Ihre Lieferverpflichtung, Ihren Anfangsbestand, Ihren Bedarf an ME des Gutes X sowie Ihre finanzielle Grundausstattung entnehmen:

Periode	1	2	3	4	5
Lieferverpflichtung	200 ME	200 ME	200 ME	200 ME	200 ME
Anfangsbestand	160 ME	140 ME	120 ME	100 ME	80 ME
Bedarf	40 ME	60 ME	80 ME	100 ME	120 ME
Finanz. Grundausstattung	800 GE	1200 GE	1600 GE	2000 GE	2400 GE

Ihre Lieferverpflichtung beträgt in jeder Periode 200 ME des Gutes X.
Ihr Anfangsbestand in der ersten Periode beträgt 160 ME des Gutes X. Somit ist Ihr Bedarf 40 ME. In jeder weiteren Periode ist Ihr Anfangsbestand um 20 ME niedriger als in der Vorperiode und somit erhöht sich in jeder weiteren Periode Ihr Bedarf um 20 ME im Vergleich zur Vorperiode.
Ihre finanzielle Grundausstattung beträgt in der ersten Periode 800 GE und ist in jeder weiteren Periode 400 GE höher als in der Vorperiode.

Ablauf einer Periode

1. In einer Auktion können Sie ME des Gutes X kaufen.

2. Danach können Sie in einem Handel ME des Gutes X kaufen bzw. verkaufen.

3. Am Ende stellen Sie in Eigenproduktion die noch benötigten ME des Gutes X selbst her, zu einem Ihnen bekannten, individuellen Produktionspreis.

Im Folgenden werden die einzelnen Punkte detailliert beschrieben.

1. Auktion

In jeder Periode werden 150 ME des Gutes X in einer Auktion versteigert (Auktionsangebot). Dadurch haben Sie die Möglichkeit, zusätzliche ME zu erwerben.
Die Auktion beginnt mit dem Preis $p = 1$ GE je ME. Zu diesem Preis können Sie ein Kaufgebot abgeben. Das Gebot besteht aus einer Menge, die angibt, wie viele ME von X Sie zu dem vorgegebenen Preis p kaufen möchten. Die Summe aller von den Teilnehmern abgegebenen Kaufgebote wird als Gesamtnachfrage bezeichnet. Ist die Gesamtnachfrage nicht größer als das Auktionsangebot, so endet die Auktion. Übersteigt die Gesamtnachfrage das Auktionsangebot, wird der Preis p um 1 GE erhöht und es startet eine neue Auktionsrunde, in der Sie erneut ein Gebot abgeben können. Dies wiederholt sich so

B. APPENDIX TO CHAPTER 3

lange bis die Gesamtnachfrage das Auktionsangebot nicht mehr übersteigt. Dann endet die Auktion und es wird jede ME des Auktionsangebots zum einheitlichen Zuschlagspreis p^* verkauft.

Der Zuschlagspreis p^* ist entweder der Preis der a) letzten oder der b) vorletzten Auktionsrunde, je nachdem ob in der letzten Auktionsrunde die Gesamtnachfrage a) gleich oder b) kleiner dem Auktionsangebot war. Im Fall a) wird Ihr zuletzt abgegebenes Kaufgebot in voller Höhe bedient. Im Fall b) wird die Menge des Gutes X, die Sie erhalten, durch Ihr vorletztes Kaufgebot bestimmt. Dieses wird, abhängig vom überschüssigen Auktionsangebot, vollständig oder teilweise bedient.

Bei der Abgabe Ihres Kaufgebotes ist folgendes zu beachten:

- Als Menge ist nur ein positiver, ganzzahliger Wert zugelassen.

- Von einer zur nächsten Auktionsrunde dürfen Sie Ihre nachgefragte Menge nach Gut X in Ihrem Kaufgebot nicht erhöhen. Bitte beachten Sie, sollten Sie in einer Runde kein Gebot abgeben, dürfen Sie in der aktuellen Auktion nicht mehr mitbieten.

- Ihr Kaufgebot darf das Auktionsangebot, also 150 ME, nicht übersteigen.

2. Handel

Im Anschluss an die Auktion können die Teilnehmer das Gut X untereinander handeln, d.h. Sie können ME des Gutes X von anderen Teilnehmern kaufen bzw. an andere Teilnehmer verkaufen. Der Handel läuft analog zu der zuvor beschriebenen Auktion ab. Im Handel können allerdings zusätzlich zu den Kaufgeboten auch Verkaufsgebote eingereicht werden. Der Handel beginnt mit dem Preis $p = 1$ GE je ME. Zu diesem Preis können Sie entweder ein Kaufgebot oder ein Verkaufsgebot abgeben.

Wie zuvor geben Sie mit Ihrem Kaufgebot an, wie viele ME von dem Gut X Sie zu dem vorgegebenen Preis p kaufen möchten. Die Summe aller von den Teilnehmern abgegebenen Kaufgebote wird als Gesamtnachfrage bezeichnet. Dementsprechend geben Sie mit Ihrem Verkaufsgebot an, wie viele ME von X aus Ihrem aktuellen Bestand Sie zu dem vorgegebenen Preis p verkaufen möchten. Die Summe aller von den Teilnehmern abgegebenen Verkaufsgebote wird als Gesamtangebot bezeichnet.

Ist die Gesamtnachfrage nicht größer als das Gesamtangebot, so endet der Handel. Übersteigt die Gesamtnachfrage das Gesamtangebot, wird der Preis p um 1 GE erhöht und es startet eine neue Handelsrunde, in der Sie erneut ein Gebot abgeben können. Dies wiederholt sich so lange, bis die Gesamtnachfrage das Gesamtangebot nicht mehr übersteigt. Dann endet der Handel und es wird jede ME des Gesamtangebots zum einheitlichen Handelspreis p^* verkauft.

Der Handelspreis p^* ermittelt sich analog dem Zuschlagspreis in der Auktion. D.h. p^* ist entweder der Preis der letzten oder der vorletzten Handelsrunde, je nachdem ob in der letzten Handelsrunde die Gesamtnachfrage gleich oder kleiner dem Gesamtangebot war. Im zweiten Fall wird, wie in der Auktion, Ihre gekaufte bzw. verkaufte Menge durch Ihr vorletztes Kauf- bzw. Verkaufsgebot bestimmt.

Bei der Abgabe Ihres Kauf- bzw. Verkaufsgebotes ist folgendes zu beachten:

1. Als Menge ist nur ein positiver, ganzzahliger Wert zugelassen.
2. In jeder Handelsrunde dürfen Sie entweder ein Kauf- oder Verkaufsgebot abgeben.
3. Für ein Kaufgebot gilt:

    - Von einer zur nächsten Handelsrunde dürfen Sie die nachgefragte Menge nach Gut X in Ihrem Kaufgebot nicht erhöhen.
    - Ihr Kaufgebot darf 200 ME nicht übersteigen.
    - Während des Handels ist es möglich, von der Käuferposition in die Verkäuferposition zu wechseln. Umgekehrt ist dies nicht möglich.

4. Für ein Verkaufsgebot gilt:

    - Von einer zur nächsten Handelsrunde dürfen Sie die Menge in Ihrem Verkaufsgebot nicht verringern. Das bedeutet natürlich auch, dass Sie in dieser Periode nicht mehr aus dem Handel aussteigen können.
    - Ihr Verkaufsgebot darf ihren momentanen Bestand an ME des Gutes X nicht übersteigen, d.h. Sie dürfen nicht mehr ME anbieten als Sie besitzen.
    - Während dem Handel ist es nicht möglich, von der Verkäuferposition in die Käuferposition zu wechseln.
    - Die Verkäuferposition können sie während dem Handel immer einnehmen, auch wenn Sie in den vorangegangenen Handelsrunden nicht aktiv waren.

3. Eigenproduktion

Am Ende einer jeden Periode müssen Sie 200 ME des Gutes X ausliefern. Falls Sie dafür nach dem Handel nicht genügend ME des Gutes X besitzen, produzieren Sie automatisch die fehlenden ME zu einem fest vorgegebenen individuellen Produktionspreis q [GE je ME] selbst. Ihre Produktionskapazität für das Gut X beträgt 200 ME. Das heißt, dass Sie jede von Ihnen benötigte Menge zwischen 1 und 200 ME des Gutes X zu dem fest vorgegeben Preis q je ME selbst produzieren können. Somit haben Sie also auch die Möglichkeit, ihre Lieferverpflichtung zu erfüllen, ohne an der Auktion bzw. am Handel teilzunehmen.
Bitte beachten Sie, in jeder Periode besitzt jeder Teilnehmer in Ihrer Gruppe einen anderen Produktionspreis q, wobei die unterschiedlichen Preise aus der Menge 3,6,9,12,15,18 [GE je ME] gezogen werden. Ihr Preis q wird Ihnen zu Beginn einer jeden Periode auf Ihrem Bildschirm bekannt gegeben. Ihr Preis q kann sich jedoch von Periode zu Periode ändern.

B. APPENDIX TO CHAPTER 3   123

        Ihre finanzielle Grundausstattung in dieser Periode
- Ausgaben für die in der Auktion erworbenen ME des Gutes X
- Ausgaben für den Einkauf von ME des Gutes X beim Handel
+ Einnahmen aus dem Verkauf von ME des Gutes X beim Handel
- Produktionskosten

Auszahlung

Ihr Periodenergebnis berechnet sich wie folgt:
Ihr Gesamtgewinn berechnet sich aus der Summe Ihrer Periodenergebnisse aus allen 5 Perioden. Ihr Gesamtgewinn wird in Euro umgerechnet und Ihnen im Anschluss an das Experiment bar ausbezahlt.

Zusammenfassung

Zusammenfassend sei nun noch einmal kurz der gesamte Ablauf beschrieben:

- Ihr Unternehmen hat in jeder der 5 Perioden eine Lieferverpflichtung von 200 ME.
- Zu Beginn jeder Periode verfügen Sie über einen Anfangsbestand des Gutes X sowie über eine finanzielle Grundausstattung.
- Der Ablauf einer jeden Periode ist wie folgt:
    1. Zuerst wird eine Auktion durchgeführt, in der Sie ME von X erwerben können.
    2. Im Anschluss daran findet ein Handel statt, bei dem die ME des Gutes X unter den Teilnehmern gehandelt werden können.
    3. In Eigenproduktion produzieren Sie die noch benötigten ME des Gutes X zu einem vorgegebenen Preis q je ME selbst.
- Von Ihnen in einer Periode zu viel erworbene ME des Gutes X werden für Sie wertlos.
- Am Ende wird Ihnen Ihr Gesamtgewinn bar ausbezahlt.

Bevor das Experiment beginnt, werden Ihnen auf dem Bildschirm einige Fragen zu den Regeln gestellt. Damit möchten wir sicher gehen, dass alle Teilnehmer die Anleitung verstanden haben.

Bibliography

Ackert, L. F. and Church, B. K.: 2001, The Effects of Subject Pool and Design Experience on Rationality in Experimental Asset Markets, Journal of Behavioral Finance 2(1), 6–28.

Anger, N.: 2008, Emission Trading Beyond Europe: Linking Schemes in a Post-Kyoto World, Energy Economics 30(4), 2028–2049.

Bagehot, W.: 1971, The only game in town, Financial Analysts Journal 22, 12–14.

Baillie, R. T., Booth, G., Tse, Y. and Zabotina, T.: 2002, Price Discovery and Common Factor Models, Journal of Financial Markets 5, 309–321.

Benz, E. and Ehrhart, K.-M.: 2008, The Initial Allocation of CO_2 Emissions Allowances: An Experimental and Theoretical Study, Working Paper, Bonn Graduate School of Economics.

Benz, E. and Klar, J.: 2008, Price Discovery and Liquidity in the European CO_2 Futures Market: An Intraday Analysis, Working Paper, Bonn Graduate School of Economics.

Benz, E. and Trück, S.: 2006, CO_2 Emissions Allowance Trading in Europe - Specifying a new Class of Assets, Problems and Perspectives in Management 3.

Benz, E. and Trück, S.: 2008, Modeling the Price Dynamics of CO_2 Emissions Allowances, forthcoming in Energy Economics .

Böhringer, C. and Lange, A.: 2005, Economic Implications of Alternative Allocation Schemes for Emission Allowances, Scandinavian Journal of Economics 107(3).

Biais, B., Glosten, L. and Spatt, C.: 2005, Market Microstructure: A Survey of Microfoundations, Empirical Results, and Policy Implications, Journal of Financial Markets (8), 217–264.

Bierbrauer, M., Trück, S. and Weron, R.: 2004, Modeling Electricity Prices with Regime Switching Models, Lecture Notes on Computer Science 3039, 859–867.

BMWT: 2006, Turbomaschinen: Wie große Energiemengen auf kleinstem Raum umgewandelt werden, Bundesministerium für Wirtschaft und Technologie .

Bollerslev, T.: 1986, Generalized autoregressive conditional heteroscedasticity, Journal of Econometrics 31(1), 34–105.

Brooks, C., Burke, S. and G., P.: 2001, Benchmarks and the Accuracy of GARCH Model Estimation, International Journal of Forecasting 17, 45–56.

Burtraw, D.: 1996, Cost Savings Sans Allowance Trades? Evaluating the SO_2 Emissions Trading Program to Date, Discussion Paper 95-30-REV .

Burtraw, D., P. K. L. B. R. and Paul, A.: 2002, The Effect on Asset Values of the Allocation of Carbon Dioxide Emission Allowances, The Electricity Journal .

Carlson, C., Burtraw, D., Cropper, M. and Palmer, K. L.: 2000, Sulfur Dioxide Control by Electric Utilities: What are the gains from trade?, Journal of Political Economy 108, 1292–1326.

Cason, T. N.: 1993, Seller Incentive Properties of EPA's Emission Trading Auction, Journal of Environmental Economics and Management 25, 177–195.

Cason, T. N.: 1995, An Experimental Investigation of the Seller Incentives in EPA's Emission Trading Auction, American Economic Review 85, 905–922.

Cason, T. N., Elliott, S. R. and Van Boening, M. R.: 1999, Speculation in Experimental Markets for Emission Permits, Research in Experimental Economics 7, 93–119.

BIBLIOGRAPHY

Cason, T. N. and Plott, C. R.: 1996, EPA's New Emissions Trading Mechanism: A Laboratory Evaluation, Journal of Environmental Economics and Management 30, 133–160.

Christoffersen, P.: 1998, Evaluating Interval Forecasts, International Economic Review 39(4), 841–862.

Christoffersen, P. and Diebold, F.: 2000, How relevant is Volatility Forecasting for Financial Risk Management, Review of Economics and Statistics 82, 12–22.

Cramton, P. and Kerr, S.: 2002, Tradable carbon allowance auctions, Energy Policy 30, 333–345.

Crnkovic, C. and Drachman, J.: 1996, A universal Tool to discriminate among Risk Measurement Techniques, Risk 9, 138–143.

Cronshaw, M. B. and Brown-Kruse, J.: 1999, An Experimental Analysis of Emission Permits with Banking and the Clean Air Act Amendments of 1990, Research in Experimental Economics 7, 1–24.

Dales, J.: 1968, Pollution, Property and Prices: An Essay in Policy-Making and Economics, University of Toronto Press.

Daskalakis, G. and Markellos, R.: 2007, Are the European Carbon Markets Efficient?, Working paper.

Daskalakis, G., Psychoyios, D. and Markellos, R. N.: 2006, Modeling CO_2 Emission Allowance Prices and Derivatives: Evidence from the EEX, Working paper.

De Jong, F.: 2002, Measures of Contributions to Price Discovery: A Comparison, Journal of Financial Markets 5, 323–327.

Dempster, A., Laird, N. and Rubin, D.: 1977, Maximum Likelihood from Incomplete Data via the EM Algorithm, Journal of the Royal Statistical Society 39, 1–38.

Diebold, F., Gunther, T. and Tay, A.: 1998, Evaluating Density Forecasts with Applications to Financial Risk Management, International Economic Review 39, 863–883.

Duan, J.-C.: 1995, The GARCH Option Pricing Model, Mathematical Finance 5, 333–345.

Ehrhart, K.-M.: 2001, European Central Bank Operations: Experimental Investigation of the Fixed Rate Tender, Journal of International Money and Finance 20, 871–893.

Ehrhart, K.-M., Hoppe, C. and Schleich, J.: 2006, Banking and Distribution of Allowances: An Experimental Exploration of EU Emissions Trading, Working paper, University of Karlsruhe.

Ellerman, A. and Montero, J.-P.: 1998, The declining Trend in Sulfur Dioxide Emissions: Implications for Allowance Prices, Journal of Environmental Economics and Management 36, 26–45.

Engle, R.: 1982, Autoregressive conditional heteroscedasticity with estimates of the variance of United Kingdom inflation, Econometrica 50, 987–1007.

Engle, R. and Granger, C.: 1987, Co-integration and error correction: Representation, estimation, and testing, Econometrica 55(2), 251–276.

Ethier, R. and Mount, T.: 1998, Estimating the Volatility of Spot Prices in Restructured Electricity Markets and the Implications for Option Values, Working paper, Cornell University.

Evans and Peck: 2007, Further Definition of the Auction Proposals in the National Emissions Trading Taskforce (nett): http://www.cabinet.nsw.gov.au.

Falk, A. and Huffman, D.: 2007, Studying Labor Market Institutions in the Lab: Minimum Wages, Employment Protection and Workfare, Journal of Theoretical and Institutional Economics 163(1).

Fichtner, W.: 2005, Emissionsrechten, Energie und Produktions, Erich Schmidt Verlag.

Franses, P. and van Dijk, D.: 2000, Non-Linear Time Series Models in Empirical Finance, Cambridge University Press.

Friedman, D.: 1991, A simple Testable Model of Double Auction Markets, Journal of Economic Behavior and Organization 15.

Garcia, R., Contreras, J., van Akkeren, M. and Garcia, J.: 2005, A garch forecasting model to predict day-ahead electricity prices, IEEE Transactions on Power Systems 20(2), 867–874.

Glosten, L. and Harris, L.: 1988, Estimating the Components of the Bid/Ask Spread, Journal of Financial Economics 21, 123–142.

Glosten, L. and Milgrom, P.: 1985, Bid, Ask and Transaction Prices in a Specialist Market with Heterogeneously Informed Traders, Journal of Financial Economics 14, 71–100.

Godby, R. W., Mestelman, S., Muller, R. A. and Welland, J. D.: 1997, Emissions Trading with Shares and Coupons when Control over Discharges is Uncertain, Journal of Environmental Economics and Management 32, 359–381.

Goldfeld, S. and Quandt, R.: 1973, A Markov model for switching regressions, Journal of Econometrics 1, 3–16.

Gonzalo, J. and Granger, C.: 1995, Estimation of Common Long-Memory Components in Cointegrated Systems, Journal of Business and Economic Statistics 13, 27–35.

Granger, C. W. J.: 1981, Some properties of time series data and their use in econometric model specification, Journal of Econometrics pp. 121–130.

Granger, C. W. J.: 1983, Co-integrated variables and error-correcting models. UCSD Discussion Paper 83-13.

Haldrup, N. and Nielsen, M.: 2004, A regime switching long memory model for electricity prices, Working Paper 2004-2, Department of Economics, University of Aarhus.

Hamilton, J.: 1989, A New Approach to the Economic Analysis of Nonstationarity Time Series and the Business Cycle, Econometrica 57, 357–384.

Hamilton, J.: 1990, Analysis of Time Series subject to Changes in Regime, Journal of Econometrics 45, 39–70.

Hamilton, J.: 1994, Time Series Analysis, Princeton University Press.

Harris, F. d., McInish, T., Shoesmith, G. and Wood, R.: 1995, Cointegration, error correction, and price discovery on informationally linked security markets, Journal of Financial and Quantitative Analysis 30, 563–579.

Hasbrouck, J.: 1995, One security, Many Markets: Determining the Contributions to Price Discovery, The Journal of Finance 50(4), 1175–1199.

Holt, C. A.: 1989, The Exercise of Market Power in Laboratory Experiments, Journal of Law and Economics 32(2), 107–130.

Holt, C., Shobe, W., Burtraw, D., Palmer, K. and Goeree, J.: 2007, Auction Design for Selling CO_2 Emission Permits Under the Regional Greenhouse Gas Initiative, Technical report, Final Report: http://www.rggi.org/docs/rggi_auction_final.pdf.

Huang, R. and Stoll, H.: 1997, The components of the bid-ask spread: A general approach, Review of Financial Studies 10, 995–1034.

Huisman, R. and De Jong, C.: 2003, Option Formulas for Mean-Reverting Power Prices with Spikes, Energy Power Risk Management 7, 12–16.

Huisman, R. and Mahieu, R.: 2001, Regime Jumps in Electricity Prices, Working paper, Rotterdam School of Management.

IETA: 2005, International Emissions Trading Association: State and Trends of the Carbon Market.

Johansen, S.: 1988, Statistical analysis of cointegration vectors, Journal of Economic Dynamics and Control 12, 231–254.

Johansen, S.: 1991, Estimation and Hypothesis Testing of Cointegrated Vectors in Gaussian Vector Autoregressive Models, Econometrica 59, 1551–1580.

Kanas, A.: 2003, Nonlinear forecasts of stock returns, Journal of Forecasting 22(4), 299–316.

Kim, C.-J. and Nelson, C.: 1999, State-Space Models with Regime-Switching: Classical and Gibbs-Sampling Approaches with Applications, Cambridge, MIT Press.

Kim, C.-J., Piger, J. M. and Startz, R.: 2004, Estimation of Markov regime-switching regression models with endogenous switching, Working papers 2003-015, Federal Reserve Bank of St. Louis.

Klaassen, G., Nentjes, A. and Smith, M.: 2001, Testing the Dynamic Theory of Emissions Trading: Experimental Evidence for Global Carbon Trading, Interim report (ir-01-063).

Kling, C. and Rubin, J.: 1997, Bankable permits for the control of environmental pollution, Journal of Public Economics 64, 101–115.

Kosobud, R., Stokes, H., Tallarico, C. and Scott, B.: 2005, Valuing tradable private rights to pollute the public's air, Review of Accounting and Finance 4, 50–71.

Madhavan, A.: 2000, Market microstructure: A survey, Journal of Financial Markets 3, 205–258.

Madhavan, A., Richardson, M. and Roomans, M.: 1997, Why Do Security Prices change? A Transaction-Level Analysis of NYSE Stocks, The Review of Financial Studies 10(4), 1035–1064.

Martens, M.: 1998, Price Discovery in High and Low Volatility Periods: Open Outcry Versus Electronic Trading, Journal of International Financial Markets, Institutions and Money 8, 243–260.

Mestelman, S., Moir, R. and Muller, A.: 1999, A Laboratory Test of a Canadian Proposal for an Emissions Trading Program, Research in Experimental Economics 7.

Milgrom, P.: 1981, Rational Expectations, Information Acquisition, and Competitive Bidding, Econometrica 49, 921–943.

Milunovich, G. and Joyeux, R.: 2007, Market Efficiency and Price Discovery in the EU Carbon Futures Market, Working paper, Division of Economic and Financial Studies, Macquarie University.

Montgomery, W.: 1972, Markets in licences and efficient Pollution Control Programs, Journal of Economic Theory 5, 395–418.

Morana, C.: 2001, A Semiparametric Approach to Short-term Oil Price Forecasting, Energy Economics 23(3), 325–338.

Mugele, C., Rachev, S. and Trück, S.: 2005, Stable Modeling of Different European Power Markets, Investment Management and Financial Innovations 3.

Muller, R. A. and Mestelman, S.: 1994, Emission Trading with Shares and Coupons: A Laboratory Experiment, The Energy Journal 15(2), 195–211.

Muller, R. and Mestelman, S.: 1998, What Have We Learned from Emissions Trading Experiments?, Managerial and Decision Economics 19(4/5), 225–238.

Pagan, A. R.: 1996, The Econometrics of Financial Markets, Journal of Empirical Finance 3, 15–102.

Paolella, M. and Taschini, L.: 2006, An econometric analysis of emission trading allowances, Working Paper, Swiss Banking Institute.

Pesendorfer, W. and Swinkels, J.: 1997, The Loser's Curse and Information Aggregation in Common Value Auctions, Econometrica 65, 1247–1281.

Plott, C. R.: 1983, Externalities and Corrective Policies in Environmental Markets, The Economic Journal 93(369), 106–127.

PointCarbon: 2004, Special issues - what determines the price of carbon, Carbon Market Analyst .

PointCarbon: 2005, Carbonmarket daily: Steel companies hesistant to offload EUA, http://www.pointcarbon.de (08.07.2005) .

Porter, D., Rassenti, S., Shobe, W., Smith, V. and Winn, A.: 2006, The Design, Testing, Implementation of Virginia's NO_x Allowance Auction, Journal of Economic Behavior and Organization .

Ramirez, O. and Fadiga, M.: 2003, Forecasting Agricultural Commodity Prices with Asymmetric Error GARCH Models, Journal of Agricultural and Resource Economics 8.

Rezek, J.: 1999, Shadow prices of sulfur dioxide allowances in Phase I Electric Utilities. Annual meeting of the American Agricultural Economics Association.

Rosenblatt, M.: 1952, Remarks on a Multivariate Transformation, Annals of Mathematical Statistics 23, 470–472.

Rubin, J.: 1996, A model of intertemporal emission trading, banking, and borrowing, Journal of Environmental Economics and Management 31, 269–286.

Schaller, H. and van Norden, S.: 1997, Regime switching in stock market returns, Applied Financial Economics, Taylor and Francis Journals 7(2), 177–191.

Schennach, S.: 2000, The Economics of Pollution Permit Banking in the Context of Title IV of the 1990 Clean Air Act Amendment, Journal of Environmental Economics and Management 40, 189–210.

Schleich, J., Ehrhart, K.-M., Hoppe, C. and Seifert, S.: 2006, Banning Banking in EU Emissions Trading?, Energy Policy 34(1), 112–120.

Schmalensee, R., Joskow, P., Ellerman, Montero, J.-P. and Baily, E.: 1998, An Interim Evaluation of Sulfur Dioxide Emissions Trading, Journal of Economic Perspectives 34, 53–68.

Schwartz, E.: 1997, The stochastic behavior of commodity prices: Implications for valuation and hedging, The Journal of Finance 52, 923–973.

Schwarz, T. and Szakmary, A.: 1994, Price Discovery in Petroleum Markets: Arbitrage, Cointegration, and the Time Interval of Analysis., Journal of Futures Markets 14, 147–167.

Seifert, J., Uhrig-Homburg, M. and Wagner, M.: 2006, Dynamic Behavior of CO_2 Spot Prices: Theory and Empirical Evidence, Working paper, Chair of Financial Engineering and Derivatives Universität Karlsruhe (TH).

Seifert, J., Uhrig-Homburg, M. and Wagner, M.: 2008, Dynamic behavior of CO_2 spot prices, Journal of Environmental Economics and Management .

Spectron: 2006, http://www.spectrongroup.com .

Stavins, R. N.: 1994, Transaction Costs and Tradeable Permits, Journal of Environmental Economics and Management 29, 133–148.

Stock, J. and Watson, M.: 1988, Testing for Common Trends, Journal of the American Statistical Association 83(404), 1097–1107.

Taylor, S.: 1986, Modelling Financial Time Series, Wiley, New York.

Theissen, E.: 2002, Price discovery in floor and screen trading systems, Journal of Empirical Finance 9, 455–477.

Tietenberg, T.: 1990, Economic Instruments for Environmental Regulation, W.W.Norton&Co, New York, chapter 16.

Uhrig-Homburg, M. and Wagner, M.: 2006, Success Chances and Optimal Design of Derivatives on CO_2 Emission Certificates, Working paper, University of Karlsruhe.

Uhrig-Homburg, M. and Wagner, M.: 2007, Futures Price Dynamics of CO_2 Emission Certificates – An Empirical Analysis, Working paper, Chair of Financial Engineering and Derivatives, University of Karlsruhe.

Ulreich, S.: 2005, Der Emissionshandel in der EU-25: Erste Erfahrungen mit einem neuen Instrument, Zeitschrift für Energiewirtschaft 29, 279–288.

Weron, R., Bierbrauer, M. and Trück, S.: 2004, Modeling electricity prices: Jump diffusion and regime switching, Physica A 336, 39–48.

Zuteilungsgesetz: 2007, Gesetz über den nationalen Zuteilungsplan für Treibhausgas-Emissionsberechtigungen in der Zuteilungsperiode 2008-2012.

VDM Verlagsservicegesellschaft mbH

Die VDM Verlagsservicegesellschaft sucht für wissenschaftliche Verlage abgeschlossene und herausragende

Dissertationen, Habilitationen, Diplomarbeiten, Master Theses, Magisterarbeiten usw.

für die kostenlose Publikation als Fachbuch.

Sie verfügen über eine Arbeit, die hohen inhaltlichen und formalen Ansprüchen genügt, und haben Interesse an einer honorarvergüteten Publikation?

Dann senden Sie bitte erste Informationen über sich und Ihre Arbeit per Email an *info@vdm-vsg.de*.

Sie erhalten kurzfristig unser Feedback!

VDM Verlagsservicegesellschaft mbH
Dudweiler Landstr. 99　　　　　　Telefon +49 681 3720 174
D - 66123 Saarbrücken　　　　　　Fax　　　+49 681 3720 1749
www.vdm-vsg.de

Die VDM Verlagsservicegesellschaft mbH vertritt

Printed by Books on Demand GmbH, Norderstedt / Germany